East Africa

New and Future Titles in the
Indigenous Peoples of Africa Series Include:

East Africa
West Africa

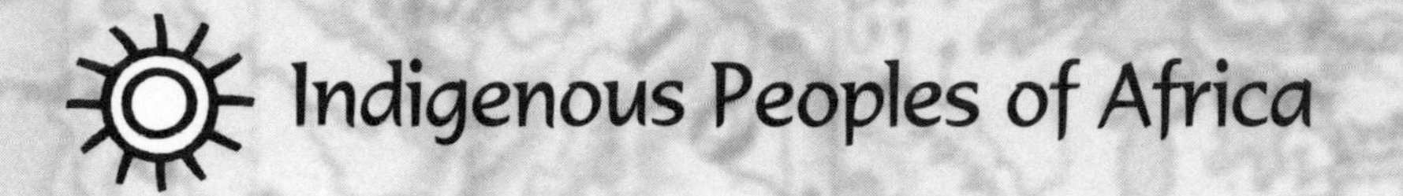

East Africa

Cynthia L. Jenson-Elliott

Lucent Books, Inc.
an imprint of The Gale Group
10911 Technology Place, San Diego, California 92127

On cover: Six East African boys in native dress.

Library of Congress Cataloging-in-Publication Data

Jenson-Elliott, Cynthia L.
East Africa / by Cynthia L. Jenson-Elliott.
p. cm. — (Indigenous peoples of Africa)
Includes bibliographical references and index.
Summary: Covers lifestyles of East African ethnic groups, Arab and European influences, religion, culture, and current problems facing East Africa.
ISBN 1-56006-969-4 (hardback : alk. paper)
1. Ethnology—Africa, Eastern—Juvenile literature. 2. Human geography—Africa, Eastern—Juvenile literature. 3. Africa, Eastern—Juvenile literature. [1. Ethnology—Africa, Eastern. 2. Africa, Eastern.] I. Title. II. Series.
DT365.42 .J46 2002
305.8'009676—dc21

2001005164

10911 Technology Place, San Diego, California 92127

Printed in the U.S.A.

Contents

Foreword

Long recognized as the birthplace of humankind, the continent of Africa has, for centuries, been inhabited by a diverse population. Physically separated by deserts, valleys, and lush forests, the people of Africa succeeded in creating unique cultural identities and lifestyles that perfectly suited the lands on which they lived. The Maasai of East Africa's Great Rift Valley, for instance, became skilled pastoralists, using young warriors to build and protect large herds of cattle and goats on the arid plains of the east. And the Ibo of Nigeria adapted their clothing and shelter-building techniques to suit life in a tropical climate.

These isolated cultures collided with outside influences sometime before A.D. 100 as Arab traders landed on African shores, and again during the fifteenth century with the arrival of Europeans. The traders came to Africa in search of valuables: gold, ivory, and diamonds. They found these items and more. One of the continent's most profitable resources turned out to be the Africans themselves. Thus began the international slave trade, which dispersed the Africans to countries around the world.

During the five centuries that followed, Africa's population was indelibly influenced by the traders and their descendants. Islam and Christianity, religions of the Arabs and Europeans, merged with traditional African beliefs. Furthermore, the power and influence of the traders—the Europeans in particular—supplanted local tribal law and led to hundreds of years of imperial rule. Yet, in spite of these influences and changes, the people of Africa managed to sustain their individual cultures and ways of life. Languages, rites of passage, tribal legends—all remained unique to the tribes that practiced them.

The *Indigenous Peoples of Africa* series examines that diversity by presenting a complex and realistic picture of the various tribal cultures. Each book in the series offers historical perspectives as well as a view of contemporary life in all of the continent's regions. The series examines family life, spirituality, art, interaction with outsiders, work, education, and the challenges faced by Africa's population today.

In many cases, those challenges are daunting. AIDS and other infectious diseases wipe out entire villages. Many African children never attend school. Human rights violations abound. Refugees of tribal warfare starve in substandard camps. Government censorship prevents citizens and journalists from speaking out against corrupt political leaders. However, even on this continent devastated by famine, ravaged by disease, and torn by war, the African people endure, bound by tradition and guided by history.

Africans also catch glimpses of a bright future. In western Africa, twenty-first century political leaders are endorsing democratic forms of government. In Kenya, a mobile library brings books to people living in isolated rural regions. And in Ethiopia, the current government sponsors training programs aimed at teaching the local population farming and agricultural techniques.

The Indigenous Peoples of Africa series attempts to capture both the Africans' history and their future, their rich culture and their current challenges. Fully documented primary and secondary source quotations enliven the text. Sidebars highlight events, personalities, and traditions. Bibliographies provide readers with ideas for further research. Each book in this dynamic series provides students with a wealth of information as well as launching points for further research.

A Land of Contrasts

The region of East Africa, consisting of the countries of Kenya, Tanzania, and Uganda, covers more than 682,000 square miles and runs from the Indian Ocean in the east to the Great Lakes region in the west. In the north, the region is bounded by the dry deserts of Ethiopia, Sudan, and Somalia, and to the south, by the nation of Mozambique.

East Africa contains some of Africa's most dramatic geography. The highest and lowest points on the continent are located here: Mount Kilimanjaro, at over eighteen thousand feet, and Lake Tanganyika, sinking more than five hundred feet below sea level. These contrasting elevations are the result of the breakup of side-by-side plates in the earth's crust. As the two plates have ripped and faulted, they have created both a string of volcanic mountains and high plateaus (Mount Kilimanjaro is only one of several volcanoes) and a tremendous valley, the Great Rift Valley. Lake Victoria Nyanza, the continent's largest freshwater lake and the source of the Nile River, forms an aquatic border separating the three countries.

The climate of East Africa is influenced not only by its unique geography but also by its location straddling the equator. The heat of the equatorial sun on the mile-high plateaus and steep valleys gives the region a variety of climatic zones, from the driest of deserts in the north to the wettest of rain forests on mountain foothills. East Africa's position on the Indian Ocean has also had a big impact on coastal climate and coastal culture. Monsoon winds blowing off the Indian Ocean have historically brought both rain and trade to the region, the latter borne on sailing ships from the Arabian Peninsula.

The Cradle of Humanity

Perhaps it is because of these unique geographical features that human life had its beginnings in East Africa. East Africa is considered by scientists to be the very cradle of humankind. Scientists believe that the earliest human beings evolved in East Africa about 100,000 years ago and spread out throughout the world from this base. Today, East Africa is home to more than 80 million people, with one of the highest population growth rates in the world. Some two hundred different ethnic groups, or tribes, make their home in the region, spilling across the boundaries of nations and sharing common histories, cultures, and even ancestries.

What is unique about the population of East Africa is that, despite East Africa's role as the cradle of humankind, most ethnic groups in the region are relatively recent immigrants—that is, they moved to the area within the last five thousand years. Today's tribes are the descendants of three distinct groups who migrated to East Africa between 3000 B.C. and A.D. 1000, as well as a small group thought to be descended from the earliest people. Scientists identify each early group by the language they spoke: Khoisan, Cushitic, Bantu, and Nilotic.

The Earliest Residents of East Africa

Scientists believe that the earliest residents of East Africa were seminomadic hunter-gatherers. They moved with the seasons to

Kenya's Mount Kilimanjaro, the highest point in Africa. The eighteen-thousand-foot mountain is also a volcano.

A rock painting made between 10,000 and 40,000 years ago by the hunter-gatherer Sandawe people in Tanzania. The earliest residents of East Africa are thought to be hunter-gatherers.

follow animal herds or the fruiting cycles of edible plants. Scientists believe these people spoke Khoisan languages, languages that use a number of clicking sounds. The Khoisan speakers lived off the land, not radically altering nature but using the resources they found in the environment. Most Khoisan speakers now live in southwest Africa, though two groups remain in East Africa.

The first migrants to East Africa were Cushitic-speaking pastoralists, or animal herders, who came from the highlands of Ethiopia. They arrived in the Rift Valley highlands of Kenya and Tanzania around 2000 to 3000 B.C. Called the "stone bowl culture" by scientists, they used stone tools to gather, and possibly grow, grain crops such as millet and sorghum. Cattle herding, however, was their main source of food, (primarily for milk). In fact, they are among the earliest pastoral herding cultures in the region. Other Cushitic-speaking groups migrated into northern Kenya from the sixteenth to the twentieth century. Most Cushitic-speaking tribes live in the arid north, but a few tribes thought to be descended from the stone bowl culture live in central Tanzania.

A thousand years after the Cushitic speakers migrated to East Africa, one of the most important migrations in the history of the world took place: the migration of Bantu speakers from west Africa. The Bantu migration was important not only for the sheer numbers of migrants but also because it spread Bantu languages throughout the continent of Africa. Today, Bantu languages are spoken by more people than nearly any other language in the world.

The Bantu speakers had an impact on East Africa that went beyond the introduction of their language. They also brought farming techniques and iron technology to the region. Prior to the Bantu, most East African tribes used stone implements to hunt and herd animals and gather wild plants. Iron technology and the farming lifestyle radically changed life in East Africa. Moreover, the Bantu speakers cleared forests for agriculture and, through intermarriage, absorbed many tribes in the region. As a U.N.-funded survey titled *General History of Africa* reports,

> Many of the hunters were sooner or later absorbed into Bantu society. . . .With the new technology, the magical control of the soil which thus began yielding grain foods, pots in which to cook them palatably, and iron tools and arrow-heads . . . the Bantu success and superiority were assured.[1]

Today, most ethnic groups in East Africa are the descendants of these Bantu-speaking immigrants. Bantu-speaking tribes live throughout the region, especially in the fertile highlands, farming, trading, and raising livestock.

The Nilotic Migration

While Bantu speakers moved into East Africa from the west, a third migration was taking place from the north. Speakers of the Nilotic language family, ironworking farmers and pastoralists, were moving from the low, dry areas of southern Sudan and southwestern Ethiopia into East Africa. As groups of Nilotic speakers migrated, the Nilotic language changed and divided into three distinct groups. These were known as the Eastern Paranilotes, the Southern Paranilotes, and the Western Nilotes.

Today, the descendants of the Western Nilotes are generally farmers who raise some animals and trade goods with other tribes living near Lake Victoria Nyanza. The Southern Paranilotes are primarily animal herders who also do some farming and live in the semi-arid highlands of western Kenya and Uganda. And most of the Eastern Paranilotes are pastoralists who live throughout the Rift Valley in Kenya and Tanzania.

Today's East African tribes speak variations of these original four languages. In many cases, two language groups have mixed together to form a hybrid language, a new language that contains a little bit of both. For example, Kimaa, the language of the Maasai, an Eastern Paranilotic tribe, contains many elements of the Cushitic

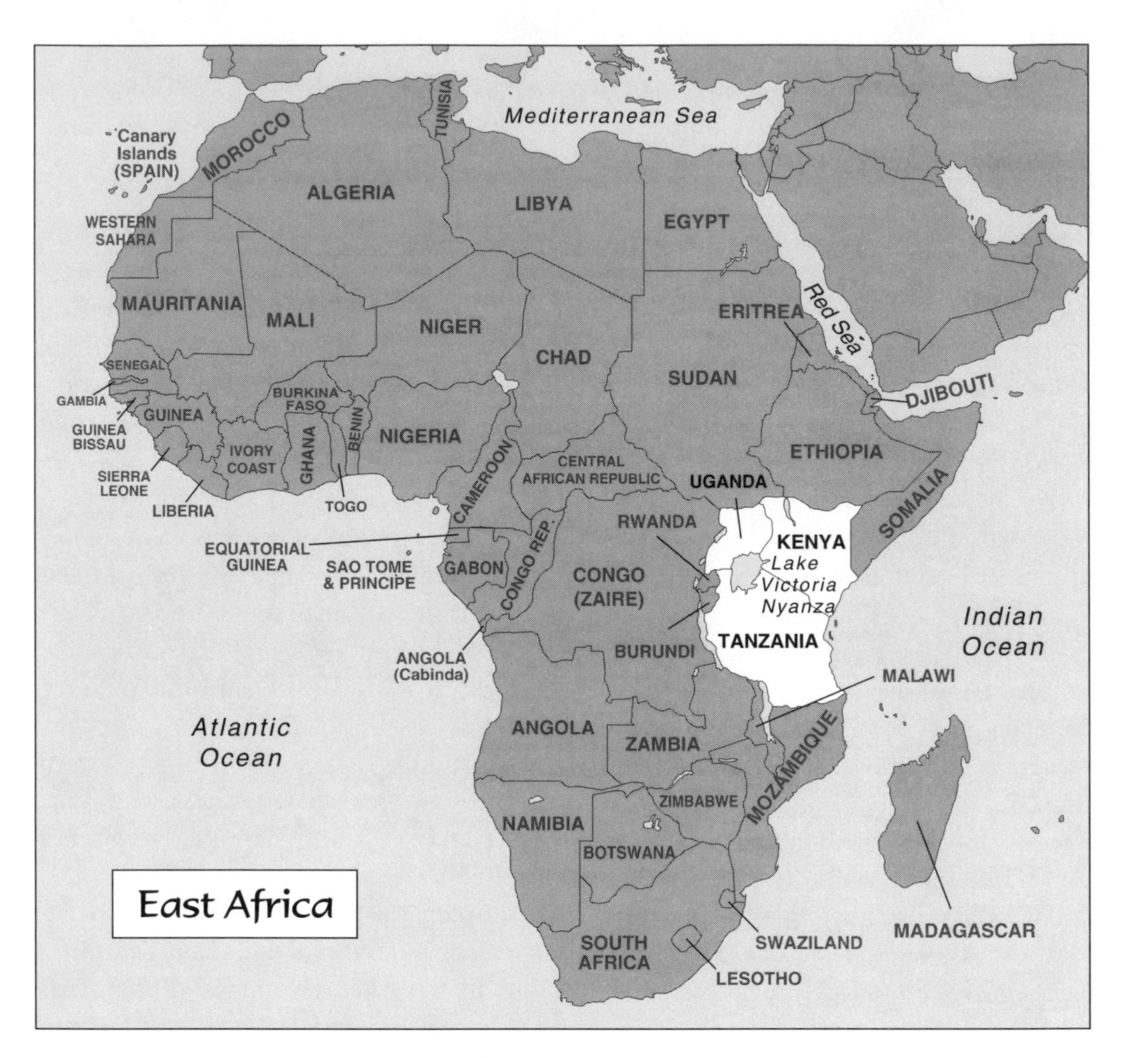

languages of neighboring tribes. And Swahili, the official language that unifies all three nations of East Africa and separates them from their lakeside neighbors Rwanda and Burundi, is a mixture of Bantu and Arabic, the language of trade on the East African coast.

In a similar way, the cultures of East African ethnic groups have mixed as well. Tribes living next to each other often intermarry and end up sharing many traditions and aspects of culture. The Kikuyu, the largest Bantu-speaking tribe in Kenya, borrowed many traditions, rites, and rituals from the Maasai, their neighbor and trading partner for many years. And many of the most important cultural practices of the Maasai were, in turn, borrowed from Cushitic speakers long before they migrated to East Africa.

Unlike the cultures of East African tribes, the lifestyles of East African ethnic groups—the ways individuals make a living and survive—depend largely on the landscape, both physical and political, in which the tribe lives. East Africans have creative and resourceful means of surviving the harsh physical and political realities of their environments. The realities of modern life include ancient problems such as disease and the harmful legacies of colonialism such as underdevelopment. To flourish in a world of such contrasts, East Africans have learned to be flexible, reaching backward to the past for stability and reaching forward to the future for hope.

Chapter 1

Lifestyles of East African Ethnic Groups

The ethnic groups that populate East Africa today live diverse lifestyles uniquely suited to the varied geography of the region. Tribes that make their home in the arid deserts of the north have nomadic lifestyles adapted for survival in this hot, harsh environment. Ethnic groups living on the rich, dark soils of mile-high volcanic mountains enjoy a lifestyle that takes advantage of the agricultural bounty of their homelands. And ethnic groups living near the Indian Ocean or on the waters of the eastern lakes region have lifestyles dependent in part on fishing.

One way to study the more than two hundred ethnic groups of the region is to group them according to lifestyle: farming, herding, fishing, or some mixture of all three. Because many areas of East Africa are suitable for more than one kind of lifestyle—for example, an area with good soil may also be near a fish-filled lake—many ethnic groups in East Africa today have mixed economies. In other words, they have more than one way of making a living.

Agriculturalists

Ethnic groups who practice pure agriculture, or farming, make up the largest lifestyle group in East Africa today. Agriculturalists are most often speakers of Bantu languages, although some Cushitic- and Nilotic-speaking tribes may also farm and keep some animals. Agriculturalists live in the areas of East Africa with the richest soils and greatest rainfall. The richest areas are the central Kenya highlands; the wet regions surrounding Lake Victoria Nyanza and other western lakes; the slopes of Mounts Kilimanjaro, Kenya, Arusha, and Meru; the lowland river deltas of Tanzania; and the narrow lowland coastal strip of Kenya and Tanzania. Many other areas of East Africa, such as the Rift Valley, are farmed as well, but with less rainfall and poorer soil than in the richer areas, inhabitants are often forced to supplement their incomes by herding animals, fishing or trade.

The Kikuyu are one tribe of Bantu-speaking agriculturalists who live in the lush rural highlands around Nairobi, the capital of Kenya. The largest ethnic group in the country, the Kikuyu have lived in central Kenya, by their own reckoning, since the beginning of time. Jomo Kenyatta, the country's first president, tells this story in his book *Facing Mount Kenya*:

According to the tribal legend, we are told that in the beginning of things, when mankind started to populate the earth, the man Gikuyu, the founder of the tribe, was called by the Mogai (the Divider of the Universe), and was given as his share the

Crops cover a hillside in a Kikuyu village in Kenya's Great Rift Valley. The agriculturalist Kikuyu are the largest ethnic group in Kenya.

> land with ravines, the rivers, the forests, the game and all the gifts that the Lord of Nature (Mogai) bestowed on mankind. At the same time Mogai made a big mountain which he called Kere-Nyaga (Mount Kenya) as his resting-place . . . and as a sign of his wonder.[2]

Kikuyu agriculturalists, like many Bantu-speaking farmers, own and farm small plots of land, or *shambas*, located near their family home, and they often live with extended families in enclosures containing several houses. Most modern Kikuyu homes are made of concrete blocks with corrugated metal roofs, while older, more traditional homes are thatched-roof buildings of wattle-and-daub construction—wood frames patched together with plaster made of mud, dung, and straw. Because the Kikuyu highlands are close to Nairobi, many Kikuyu today have electricity and running water in their homes and villages, but most agriculturalists in East Africa do not. For most farmers, water must be hauled from natural water sources such as rivers and lakes or from a village pump. Light, heat, and cooking fuel often come from oil lamps and traditional cooking fires.

Agricultural villages, including those of the Kikuyu, are usually centered along a main road containing a few small stores; government service centers such as a police station, school, or health clinic; and several churches. In their villages, the Kikuyu regularly take advantage of government services such as education, and most children attend school for some period of their lives. Adults often work outside their homes or farms, in trade or for the government or in large-scale agriculture.

Agriculturalists Who Also Keep Livestock

Agriculture alone cannot support people living in regions with sparse rainfall and moderate soil. For this reason, most agriculturalists in East Africa also keep some livestock or work in trade to round out their incomes. The Nilotic-speaking Luo are an example of such a mixed lifestyle. Like all Nilotic tribes, the Luo made a living in the distant past by herding animals. Today, the Luo live in the region northeast of Lake Victoria Nyanza, an area suitable for agriculture, fishing, and cross-border trade with Uganda.

Many Luo continue to herd cattle in areas surrounding their villages or cross the border from Kenya to Uganda for work in the black markets of the Great Lakes region. Despite this mixed economy, however, Luo life continues to revolve around a stable village center where people receive services and interact with neighbors (where their ancestors are buried). For agriculturalists with a mixed economy, agricultural land is the glue that binds the individual to the tribe, and the tribe to the past and future.

Pastoralists

Pastoralists, or animal herders, are bound by a sense of place in a different way than

Buganda: A Wealthy Agricultural Kingdom

Uganda was once the home of several large and powerful kingdoms, both agricultural and pastoral, ruled by wealthy kings. Even today, the traditional leaders of some of these ancient kingdoms hold ceremonial and spiritual power—power endorsed by the modern-day government—to rule their people. The most powerful of these kingdoms was a Bantu-speaking agricultural group, the Buganda, ruled by a king called the *kabaka*.

The Buganda people have a complex and aristocratic history made possible by the wealth produced by banana agriculture. In the sixteenth century, bands of farmers from central Uganda began joining together under chiefs to protect themselves from other strong tribes. By the eighteenth century, these bands had formed into a highly organized government with a *kabaka*. Over the next century, the *kabakas* created a bureaucracy which became politically and economically powerful because of banana agriculture.

Bananas, an import from Asia, grew easily in the rich soils surrounding Lake Victoria Nyanza, and banana cultivation resulted in economic stability and population growth. The Reverend John Roscoe, a visitor to the Buganda, wrote in his 1911 account *The Baganda: An Account of Their Native Customs and Beliefs*,

"The [bananas] grow so freely that a woman can supply the needs of her family with a minimum of labor, and with the barkcloth trees and man can supply their clothing. The country had all its needs supplied by its own products for many years."

Bananas were cultivated by landless peasants who farmed large estates owned by chiefs loyal to the *kabaka*. Peasants also paid taxes to the *kabaka* and the chiefs. Bananas grew so easily and took so little labor that the peasants were also required to perform public work, such as the building of roads. Such labor and wealth led to the creation of an impressive bureaucracy and infrastructure based in a capital city, to which all roads led.

The royal tomb of the kabaka, *the Buganda king. The* kabakas *ruled a powerful kingdom made rich from banana cultivation.*

agriculturalists are. As it is for agriculturalists, the boundaries of pastoralists' territories are set by tradition and, in modern times, law. But unlike farming tribes, individual animal herders do not own the land; they merely graze and water their animals on traditional tribal lands that are held and cared for collectively by a clan or tribal group.

A Turkana goatherd in Kenya watches over his animals. Pastoralists use Kenya's arid land to graze and water their herds.

The area of East Africa devoted to pure pastoralism is enormous. Large sectors of Kenya, so arid they have no rivers and may have no rainfall for years at a time, are suitable for no lifestyle other than the herding of camels, cattle, and goats. The number of tribes that continue to practice pastoralism, however, is small, for the land is so dry that it will support very little life. Pastoralists must lead a nomadic lifestyle, traveling from one traditional grazing land to another, from one watering hole to another, when the need arises.

In East Africa, there are two types of pure pastoralists, who lead somewhat different lives: Paranilotic-speaking pastoralists, who are seminomadic, moving two or three times a year, and Cushitic-speaking pastoralists, who move much more frequently. The lifestyles of these groups are ruled by the land on which they live. The harsher the environment, the more the nomads must move.

Gabbra: Cushitic-Speaking Pastoralists

The Gabbra are a Cushitic-speaking nomadic pastoralist group living in the northern reaches of Kenya, near

Ethiopia, in and around the Chalbi Desert, the hottest, driest area of East Africa. Water is scarce in the north. In much of the Gabbra land, there are no rivers or lakes, and the few deep and ancient wells are scattered broadly. Because of this lack of water, the Gabbra's lifestyle is built around mobility, the frequent, nomadic search for grazing land and water sources for their herds of camels and some cattle, sheep, and goats.

Gabbra homes, like the homes of other Cushitic-speaking pastoralists, reflect their need for mobility. Houses are made of bendable wooden frames interwoven with animal hides. When it is time to move, the homes are simply taken apart and packed up like tents to be transported on the backs of camels. And then, in a long caravan of camels, the Gabbra move to a new location.

Although the Gabbra move frequently, their movement is not random. It is carefully planned to take advantage of watering sources. Since many springs tend to be on the periphery of Gabbra grazing areas and in unusable desert, their grazing strategy tends to be opposite of most other pastoralists. Gabbra move to the highlands just after the rains start to take advantage of new standing pools in the Dida Galgalla plains; then as these waters dry up, they move gradually south, closer and closer toward permanent water sources in the Chalbi Desert.

Maasai

One of the Maasai's most powerfully held beliefs about cattle is that at the beginning of time *Ngai*, or God, gave all the cattle on earth to the Maasai. In the book *Maasai*, coauthors Tepilit Ole Saitoti and Carol Beckwith cite the first European account of contact with the Maasai, published by Dr. Ludwig Krapf in 1860.

"They live entirely on milk, butter, honey and the meat of black cattle, goats and sheep . . . having a great distaste for agriculture, believing that the nourishment afforded by cereals enfeebles, and is only suited to the despised tribes of the mountains. . . . When cattle fail them they make raids on the tribes which they know to be in possession of herds. They say that *Engai* (Heaven) gave them all that exists in the way of cattle and that no other nation ought to possess any. . . . They are dreaded as warriors, laying waste with fire and sword, so that the weaker tribes do not venture to resist them in the open field, but leave them in possession of their herds and seek only to save themselves by the quickest possible flight."

Maasai: Paranilotic-Speaking Pastoralists

Although the availability of water and grazing land is a limiting factor for all pastoral tribes, governing the size of their clans and their movements, not all pastoral groups need to move as often as the Gabbra in order to survive. The Maasai, East Africa's most famous pastoralists, live in the Rift Valley of Kenya and Tanzania. In the Maasai's homeland, rainfall is generally seasonal and regular, though not abundant, and the soil is too poor for large-scale agriculture. The Maasai's lifestyle and culture revolve around the herding of cattle, goats, and some camels. Most of their food comes from cattle—milk, yogurt, and occasional blood and meat. Harvard-educated Maasai tribesman Tepilit Ole Saitoti, writing with coauthor Carol Beckwith in *Maasai*, provides details:

> The Maasai drink blood during the dry season when they run short of milk. The animal is not killed in obtaining the blood, but rather the tip of an arrow is used to make an incision in its jugular vein. Warriors often drink the blood of healthy animals with the belief that it will give them strength. When a woman gives birth, when a person is wounded, or when a boy or girl is circumcised, he or she is given blood to replace the lost blood. In recent years, the custom of blood drinking has partially given way to the consumption of a cereal made by mixing cornmeal and water.[3]

The Maasai are seminomadic, moving a few times a year between two or three main grazing areas. In each area, a family will keep a semipermanent and easily repaired homestead called a kraal. Each kraal contains a simply built, thornbush

Maasai villagers in Kenya collect cattle blood, which they drink when milk is unavailable. The Maasai believe the blood gives them strength.

enclosure, or *boma*, inside of which are several small houses for members of the extended family and an inner thornbush enclosure for herd animals to sleep in at night. The thornbush *bomas* not only serve as the living space of the family and animals but also protect the herds from wild predators.

Inside the outer *boma* sits a circle of loaf-shaped wattle-and-daub houses. There are no windows or chimneys in Maasai homes, so the smoke from cooking fires in the center of the house must find its way out the single door. Women are responsible for building and maintaining these homes. As Saitoti and Beckwith explain,

> One of the first responsibilities of a newly married woman is to build her home. The basic structure is formed by weaving long tree branches together. Open spaces are patched with leaves and grass before a final coating of dung is plastered over the entire surface. . . . Maintaining the house, and especially keeping it weather tight, are also the woman's responsibilities. Just before the rainy season, she will patch the roof, filling the cracks with fresh dung.[4]

A Maasai woman stands in the door of her house. Maasai homes are made out of branches, leaves, grass, and dung.

Many Maasai women, along with their children and the elderly, spend part of the year living near a village center so that the children can have some access to school. Maasai men, especially young warriors, continue to follow the herds, disdaining agricultural products and priding themselves on eating only the products of their herds.

Though many Maasai disdain agriculture and pride themselves on being pure pastoralists, some traditionally pastoralist tribes farm small plots of land for at least part of the year, when rainfall is suitable for agriculture. For example, the Nandi, who live in highland areas from northwestern Kenya and eastern Uganda down through western-central Kenya, are mainly dependent on their herds for food, but agricultural produce helps them make an adequate and healthy living.

Fishermen and Coastal Cultures

Agriculture also rounds out the incomes of fishermen who ply the waters of the Indian Ocean and the Great Lakes region of East Africa, the long string of lakes that form an aquatic boundary in the western part of the region. Fishing cultures have lived in East Africa for as long as the climate has supported that lifestyle. Three thousand years ago, when the climate of East Africa was much wetter and the entire region was filled with lakes, many cultures grew up with fishing as a way of life. Today, no ethnic group in East Africa makes its living solely by fishing, but many use fishing as a supplemental livelihood. The Turkana, an Eastern Paranilotic ethnic group, for example, fish the waters of Lake Turkana in arid northern Kenya, but primarily herd cattle for a living. The Buganda, on the shores of Lake Victoria Nyanza, also supplement a healthy agricultural economy with fishing. And the Swahili and Zaramo people who live along the East African coast make a living from fishing, agriculture, and trade.

The Zaramo, Bantu speakers who live near the southern coast of Tanzania, are a good example of a culture that finds its income both in fishing and in other sources. Anthropologist Marja-Liisa Swantz notes in a book about the Zaramo, "More than half of the male population has fished at one time or another in their lives, and even many women go to sea at certain times of the year for netting tiny fish, called *dagaa*, with large pieces of calico or black cloth."[5]

Though the Zaramo traditionally live about five miles inland, their lifestyle is tied to the sea. Each morning at dawn, Zaramo fishermen trek to the shore. Up and down the coast, tiny boats with one or two occupants can be seen setting out to ply the coastline with fishing nets. Wearing the small lace caps and long shirts that identify them as Muslims, Zaramo fishermen make a simple living hauling fish from the sea for sale and trade in the village. Back at home, their fishing income is supplemented by agriculture or trade. Most families farm cassava, maize, and rice on small *shambas* and cultivate cashew nuts on larger collective plots shared by the village.

Hunter-Gatherers

Collective farming, in which village residents cultivate fields together and all share in the bounty of the harvest, was introduced by the Tanzanian government after independence in 1961. Many groups who have not traditionally farmed have been taught agricultural techniques through government-sponsored collective farms, and have thus been encouraged to give up their traditional lifestyles.

Two ethnic groups who traditionally made their living as hunter-gatherers (that is, by hunting wild animals and gathering wild plant food) in the forests of central Tanzania have been introduced to agriculture in this way. Making up less than 1 percent of the population of Tanzania, the Hadza and Sandawe hunter-gatherers are thought to be descended from the earliest peoples of East Africa, and they are the only remaining Khoisan speakers in the region. Khoisan is a language that includes numerous "click" sounds and is spoken mainly by the bushmen of southwest Africa.

Two Kenyan men in a village near Lake Turkana proudly display a fish. Though no ethnic groups earn a living entirely by fishing, many use it as a supplemental means of supporting themselves.

As late as 1981, the Hadza continued a nomadic lifestyle, moving every few days, hunting game animals and gathering wild plant food. In 1981, it was written that "The Hadza remain hunters and gatherers, few in number, fairly mobile and expert in finding and winning the wild food resources of their territory."[6] They were known to nearby agricultural groups as wild honey gatherers, and often traded honey for agricultural produce. For most of the year, the Hadza lived in beehive-shaped huts quickly assembled from intertwined branches covered with grass. They would set up these huts in new locations every few days.

Today, the Hadza are still known for their honey-gathering expertise and cultural traditions that are different from those of their agricultural neighbors. In the 1980s, however, the Tanzanian government encouraged the Hadza to move into collective villages, where they were taught agricultural techniques. Many gave up their nomadic lifestyle, but they did not give up all the traditions associated with hunting and gathering or their unique Khoisan "click" language.

City Life

Nearly every language in East Africa, including Khoisan languages, can be heard in the urban areas of the region. And yet, despite the diversity of cities, urban dwellers constitute a minority in East Africa; only 15 to 25 percent of the population lives in cities. In large metropolitan areas such as Nairobi and Mombasa

Nairobi, Kenya's capital and largest city. No more than one quarter of East Africans live in cities.

A densely-populated neighborhood in Mombasa, Kenya. Most East African city dwellers live in apartment buildings or small houses.

in Kenya, Dar es Salaam in Tanzania, and Kampala, Uganda, as well as numerous smaller cities, ethnic groups with diverse cultures live together in relative harmony.

Life in East African cities bears many similarities to life in urban centers worldwide, with long broad streets, skyscrapers, and modern services such as running water, electricity, and transportation.

Tribalism

In some parts of East Africa, tribalism, or identification with one's ethnicity, is considered to be both old-fashioned and a cultural and political scourge. Tanzanians, in particular, believe tribalism to be a dangerous remnant of colonialism. Tribalism was used by Europeans to pit one tribe against another, and Tanzanians believe that their nation's unity and prosperity depends on giving up such "backward" notions. To a large degree, Tanzanians identify themselves as Tanzanian citizens rather than as members of a particular tribe. In all three nations, in fact, the loosening of tribal identity and the strengthening of national identity may be an important aspect of creating strong, unified nations where the rule of law replaces traditional systems characterized by favoritism.

Housing ranges from Western-style mansions surrounded by large gardens—often a remnant of colonialism—to extensive shanty towns on the outskirts of cities, with homes built out of cardboard boxes and scrap-metal roofs. Somewhere in between live the majority of urban East Africans in tall, cement-block apartment complexes or small cement row houses.

Some urban dwellers have access to land for gardening or keeping some chickens or goats, but many do not. Maintaining traditional lifestyles is generally difficult in urban areas. Some urban East Africans, though, maintain cultural traditions such as wearing traditional clothing. Dress in East African cities tends to be varied. While some wear Western attire such as pants and shirts or skirts and dresses, others wear traditional ethnic clothing such as the red waist-wraps of Maasai men and women.

Ethnicity carries less importance in the cities than it does in the countryside, but many East Africans find ways to stay connected to their ethnic roots in the city. Some people frequent businesses or cafes owned by members of their own tribe. For others, sports teams help them stay connected. The Luo and Luyha ethnic groups, for example, which traditionally live in close proximity in the Western Province of Kenya, stay connected to their ethnic groups in Nairobi by keeping alive a keen, and occasionally violent, rivalry between Luo and Luyha soccer clubs in the city.

Ethnic Roots

Most East Africans remain near their ethnic roots for their entire lifetimes. Those who leave, however, maintain cultural connections to their homeland by frequently visiting tribal lands for celebrations and ethnic rituals and staying in contact with country relatives. Maintaining a connection to a tribal lifestyle—such as agriculture or pastoralism—often proves more difficult. Urban East Africans, in particular, by removing themselves from their agricultural or pastoral roots, often lose a viable means of survival. Even in the modern age, many see ancient tried-and-true lifestyles, connected to the land and the varied environments of the region, as the best way of supporting East Africa's burgeoning population.

Chapter 2

The Arab Influence on East Africa

The Arab influence in East Africa is evident throughout the region. From the architecture of coastal buildings and the dress of the people to the language, religion, economy, and even ethnicity of many East Africans, Arab influence is strong throughout Kenya, Tanzania, and even inland to Uganda. Since the time of their arrival in trading vessels on seasonal monsoon winds, the Arabs have quietly but powerfully altered the culture of the entire region.

Before the Arab, Persian, and Mediterranean traders came to the East African coast sometime before the year A.D. 100, there were no written records detailing life in the region. The people of East Africa had not yet developed a written language. So scientists and historians must piece together the history of the region from archaeological and linguistic evidence. From this evidence, they surmise that the East African coast prior to 100 was occupied by Bantu and Cushitic farmers and fishermen. In his book *History of Africa*, Kevin Shillington writes about the people of the coast as recorded in *The Periplus of the Erythrean Sea*, the first written record describing Azania, the East African coast, in about A.D. 100:

> The people of Azania were clearly experienced fishermen, well-practiced in the use of small boats along the coastal waters offshore. They fished and caught turtles from dugout canoes and they sailed among the islands in small coastal boats made of wooden planks knotted together with lengths of coconut fibre. Each market-town was under the rule of its own chief, though the *Periplus* [an ancient Greek sailor's guide to the Indian Ocean] tells us little more about the people except that they were dark-skinned and they were tall.[7]

Arab Traders Arrive in East Africa

Into this farming and fishing world came Arab traders looking for sources of ivory and other luxury goods. The Arabs traded ivory, tortoise shell, rhinoceros horn, and coconut oil for items the Africans desired, "iron goods, particularly lances [like spears] . . . hatchets, daggers and awls, various kinds of glass and 'a little wine and wheat.'"[8] At first the Arabs stayed close to the coast. Only later did they venture inland on trading expeditions, following the routes of their African trading partners.

From the beginning, the Arab presence in East Africa was a partnership. Arabs intermarried with African families, adopted their language and culture, and lived on as intermediaries of future Arab trading ventures. The unknown author of *The Periplus of the Erythrean Sea* explains that rulers from Himyar, a state in southwest Arabia, would

> 'send thither many large ships, using Arab captains and agents, who are familiar with the natives and intermarry with them, and know the whole coast and understand the language.' . . . [This] may thus have begun the process of creating a class of coastal sea-going and trading people of mixed parentage, who acted as local agents for the international system of trade.[9]

Over the next two thousand years, the intermarriage of Africans and Arabs created a new ethnicity and language endemic

Finding History Without a Written Record

Scientists and historians have two tools at their disposal to trace the history of East Africa before written records were established: archaeology and linguistics. Archaeology is the study of artifacts, such as pottery and tools, that a people have left behind. These artifacts tell historians much about the lifestyles of the people, what they ate, and how they lived.

Historians can also tell a great deal about the origin of a people by tracing their language development, a study called linguistics. By studying how language has spread and changed throughout the continent over time, scientists can trace the movements of people and tribes. For example, by looking at words present in all Bantu languages such as the word for "yam," scientists can guess that yams were part of the culture of the original Bantu group that migrated across the continent. They can also guess where they might have lived—in an area with a climate suitable for yam cultivation. Other early Bantu words for "fishing" and "rivercraft" indicate that these Bantu lived in an area where they made part of their living from the water. The absence of common words for things such as goats and grain crops also tells linguists that livestock and grain crops were introduced into the culture later in their migration and that the original Bantu did not cultivate grains or raise goats.

Monsoon Winds

Trade from the Arabian Peninsula to the East African coast would not have been possible without the seasonal monsoon winds that carried dhows, the Arab sailing vessels, along the coast. According to Kevin Shillington, in his book *History of Africa*,

"In the western Indian Ocean the monsoon winds blow towards east Africa between November and March and towards India and the Persian Gulf between April and October. This seasonal pattern of monsoon winds largely influenced the pattern of cross-ocean trade that developed between the east African coast and the Islamic world of western Asia. . . . The journey across the ocean would take several months. This did not leave them much time for trading along the east African coast before they had to turn for home on the southwest monsoon."

For this reason, trade between the Arab world and the East African coast did not venture much farther south than the current coast of Tanzania. The northern towns of Mogadishu, Barawa, and the Lamu islands of Kenya were the most common frequent stops for the dhows.

In addition, the relatively short time between the wind's reversal discouraged Arab traders from venturing inland to procure goods. Instead, East African tribes from the hinterland brought goods to coastal markets. Arab traders could briefly stop at these trading posts to pick up products, and then move quickly on to the next coastal settlement.

The dhow, a centuries-old sailing vessel that carried Arab traders to East Africa.

to East Africa: the Swahili people and their language, Kiswahili. And over time, the international trade begun on Arab sailing ships would link East Africa to the world economy—to its ultimate economic detriment.

The Arab Influence on the East African Coast

The Arab influence on the East African economy, culture, and lifestyle is especially apparent in the coastal regions of Kenya and Tanzania, particularly in the culture and ethnicity of the coastal Swahili people. The Swahili are a coastal ethnic group of mixed Arab, African, and Persian descent. Their ancestors first came to East Africa as traders, and later, after A.D. 700, as Shi'ite Muslim refugees who settled, traded, and intermarried with African families. The word *swahili* comes from the Arab word *sahil*, which means coast. *Swahili* means the people of the coast, and *kiswahili* means the language of the people of the coast. Kiswahili, the official language of Kenya, Tanzania, and Uganda, is a mixture of Bantu and Arabic.

Kiswahili first emerged as a language, spoken in the homes of coastal people, around 1100. Arabic, however, continued to be the language of trade, education, and religion for many hundreds of years, but in time, a written version of Kiswahili was developed. Arabic script is used to write Kiswahili; thus, Swahili culture continued to develop as a hybrid of Arabic and African influences.

Today, the physical evidence of this mixture of elements makes a strong showing on the Indian Ocean coastal strip and islands. The architecture of the region is perhaps the most obvious feature of the Arab presence. Before the year 1000, most buildings along the coast were traditional African stick-frame houses plastered with mud and dung. But as more Arab and Persian Muslims came to settle in the region and the wealth of coastal cities increased, many wealthy traders built rectangular Arab-style houses of pink coral stone. Mosques, Islamic houses of worship, were also built in many coastal towns as more and more African and Swahili rulers converted to Islam. Today, many of these early structures remain, especially on coastal islands such as Lamu and Zanzibar. Newer buildings often show an Arab influence as well.

The dress of coastal people also reveals the Arab influence on the region. Men and women in the coastal strip of Kenya and Tanzania and on the offshore islands often dress modestly, in traditional Islamic garb with an African flavor. Men often wear long dress-like shirts called *kanzu* and small lace caps on their heads. These caps are part of Islamic law which requires followers to keep their head covered in the presence of Allah. Women on the coast also dress modestly, often covering their heads and much of their faces with colorful African cloths.

In addition to the clothing and architecture of the region, the Arab influence on coastal Swahili culture can also be heard in the melodic call-to-prayer emanating from mosques five times each day. Nine percent of East Africans are Muslim, mostly living on the coast. For Muslims, daily life revolves to a great extent around a local mosque, and the Arab influence is felt most strongly in their religious life. In other parts of East Africa, Muslim missionaries continue to bring the word of Allah to disparate tribes, hoping to win converts and change lives.

While Islam has had a far-reaching effect on the lives and culture of the coastal Swahili people, it can be argued that the Arab-influenced economy of the region has had a stronger impact. Before the arrival of Arab traders, the people of the coast were largely farmers and fishermen who also participated in some trade among coastal islands and with inland tribes. With the arrival of Arab traders and the growth of Swahili culture, a whole new world of trade opened up, paving the way for new cultural influences that changed life in the coastal and inland regions forever.

The Arab influence on East Africa is seen most prominently in architecture, especially mosques, such as this one in Nairobi, Kenya.

A Swahili man (wearing a traditional Islamic lace cap) on the Kenyan island of Lamu reads the Koran, the Muslim holy book.

Trade and the Arab Influence on the Interior of East Africa

The growth of trade with the Arab world changed the economy of East Africa's coast and hinterland dramatically. Trade began on a small scale, but over time it grew to meet the appetite that Arabia, India, and the Mediterranean had for African luxury goods: gold, ivory, pearls, rhinoceros horn, ambergris (a substance from whales used to make perfume), and leopard skins and other furs. This trade fueled the growth of small city-states, and later large centralized governments, along the Swahili coast from 1100 to the late 1800s.

City-states grew in size and wealth all along the coast, fueled by a steady input of goods and labor from the inland regions of East Africa. These goods and services were often not given freely. Some coastal rulers made a practice of raiding inland tribes for luxury goods, cattle, and slaves, claiming their takings as taxes and tribute or as spoils of a "holy war," or jihad, against non-Muslims. For this reason, some coastal communities had hostile relations with inland tribes, whom the Arabs called the "Zanj." Historian Kevin Shillington writes,

> The main cause of friction were raids by the Swahili themselves into the interior in search of livestock and other booty and further captives to enslave. At the time of Ibn Battuta's [a trader to the coast] visit in 1331 the sultan of Kilwa was conducting what he claimed was a jihad against the "pagan Zanj" of the interior. The nature of his "holy war" is revealed by Ibn Battuta's comment that the sultan "frequently makes raids into the Zanj country, attacks them and carries off booty." Some of this he kept for himself, the rest he put into a special fund for entertaining foreign visitors, such as Ibn Battuta himself. The sultan was surnamed

Life in Zanj, the Swahili Coast

Before the year 1000, the Arabs called the Africans of the east coast "Zanj," and their territory, the "Land of Zanj." Al-Masudi visited the region in A.D. 916 on an Arab dhow from Oman, in southeastern Arabia. His written observations, quoted in Kevin Shillington's *History of Africa,* provide a picture of what life was like in coastal East Africa in that period.

"The land of Zanj produces wild leopard skins. The people wear them as clothes, or export them to Muslim countries. They also export tortoise-shell for making combs, for which ivory is likewise used. . . .The Zanj use the ox as a beast of burden, for they have no horses, mules or camels in their land. . . .There are many wild elephants in this land but no tame ones. The Zanj do not use them for war or anything else, but only hunt and kill them for their ivory. . . .

The Zanj have an elegant language and men who preach in it. One of their holy men will often gather a crowd and exhort his hearers to please God in their lives and to be obedient to him. He explains the punishments that follow upon disobedience, and reminds them of their ancestors and kings of old. These people have no religious law: their kings rule by custom. . . .

The Zanj eat bananas, which are as common among them as they are in India; but their staple food is millet and a plant called *kalari* which is pulled out of the earth like truffles. The also eat honey and meat. They have many islands where the coconut grows; its nuts are used as fruit by all the Zanj peoples."

Abu al-Mawahib ("the Father of Gifts") "on account of his numerous charitable gifts"![10]

Other city-states chose to develop legitimate trading relations with ethnic groups of the hinterland. Trade between inland tribes and the coast—the exchange of produce of the fields for products of the sea—had been a feature of East African life long before the arrival of the Arabs. With the growth of the Swahili city-states, however, trade expanded dramatically. Trade became a westward search for new goods and markets as well as an eastward journey to bring products to market.

Expansion of Trade Brings Many Changes

Enterprising ethnic groups of the interior aided in the early expansion of trade. In order to feed the immense international market for luxury goods and, later, slaves, some ethnic groups became middlemen, traders who brought goods to the coast, while others became primary producers who found or created the goods for market. These economic roles were new to most tribes. Some ethnic groups gave up traditional means of making a living, such as agriculture or pastoralism, in order to focus on trade. Shillington describes one such ethnic group, the Nyamwezi, a previously agricultural Bantu-speaking tribe in what is now Kenya and Tanzania:

> In the western region the Nyamwezi in particular were, by the end of the eighteenth century, organizing themselves as professional traders and ivory-porters. From south of Lake Victoria Nyanza they were well-placed to develop long-distance trading routes between the interlake kingdoms and the east African coast. The trade in ivory, and later slaves, from the interior was to become an important feature of the region.[11]

Lifestyle was not the only aspect of ethnic groups that was changed by the Arab-Swahili coastal trade. The cultures of many tribes—their values, rituals, ways of interacting with people, and religious views—changed as well.

One critical cultural change was a change in how ethnic leaders were chosen. Some ethnic groups, whose leaders were traditionally old and wise elders, began turning instead to young, wealthy traders for leadership. Wealth came to be valued over age as a sign of power and wisdom. This change would have dire consequences when the Europeans later sought to establish trading and political relations in East Africa. Seeking tribal leaders, the Europeans contacted not the traditional elders but the wealthy traders, and often made official treaties with these unofficial "chiefs."

The lure of power and wealth offered by trade with the Arabs changed ethnic groups in other ways as well. Some tribes who had been closed culturally opened up. The Miji Kenda, for example, were Bantu speakers whose unique lifestyle included living in culturally closed, fortified hillside villages called *makaya* overlooking Swahili towns. According to a U.N.-funded history of Africa, "Fortification contributed towards the development of group identity and solidarity. . . .The *makaya* became more than mere havens of security. . . . They [occupied] a central place in the socio-religious life of the Miji-Kenda."[12] Trade altered Miji Kenda culture dramatically. As individuals began to live outside Miji Kenda forts to pursue commercial activities, the unity of their cultural system was lost, and the fortified settlements of the Miji Kenda became a thing of the past.

Wealth was not the only aspect of Arab trade that changed tribes culturally, however. The simple interaction with new groups beyond their borders changed the cultures of many tribes. As ethnic groups involved in trade expanded beyond their borders and met other tribes with whom they had not had contact, new alliances developed. In some tribes it became common

The Arab Slave Trade

Long before the Europeans began trading in African slaves, Arab dhows brought East Africans to work as household slaves or in salt mines and plantations in Arabia and the Persian Gulf. For centuries, slaves were taken captive during fights and raids between inland East African tribes. If they were not used as slaves by the tribe capturing them, they were brought to the coast and sold to the Arab traders. Although the numbers of slaves sold to the Arabs were generally small, by 868 the "slaves were numerous enough at Basra [on the Persian Gulf] to rise in revolt," according to historian Kevin Shillington's *History of Africa.*

The Arab slave trade increased in the sixteenth century as French plantation owners in the Indian Ocean discovered East Africa as a source of slave labor. Arab and Swahili traders began buying slaves captured by the Yao and Nyamwezi tribes and reselling them to the Europeans. Then, in the nineteenth century, the Arab use of slaves skyrocketed, and, according to Shillington, the Arab market for slaves became enormous:

"In the middle decades of the nineteenth century there was a rapid growth in the Arab demand for slaves to man their plantations on Zanzibar and surrounding islands. . . . In the 1820's Sultan Seyyid Said encouraged Arabs to set up clove plantations on the islands of Zanzibar and Pemba. The plantations . . . were worked by slave labour from the mainland. . . . There followed a rapid increase in the slave trade from the mainland to Zanzibar and the island became the largest slave market along the east African coast.

It has been estimated that at the peak of the trade in the 1860's east Africa was exporting up to 70,000 slaves a year."

The slave trade ended in 1873 when the sultan of Zanzibar took British suggestions to heart and decided to close the slave market. By that time, slaves were incorporated into all aspects, and at all levels, of Arab coastal society.

Arab slave dealers and their slaves on the streets of Zanzibar in the nineteenth century.

to take a "foreign" wife, a wife from another culture. Through such intermarriages, important rituals were shared and cultures blended. A writer quoted in another volume of the U.N. study, citing the adoption of ritual practices as a result of the interaction between groups in eastern Kenya, summarizes: "New methods of healing and divining, of controlling rainfall, and of spirit possession spread throughout Kenya as individual cultural practices became amalgamated into regional patterns."[13]

Trade Encourages Blending of Cultures

Cultures also blended, in the new multiethnic towns that emerged along expanding caravan routes. As Swahili tradesmen began developing new inland trade routes to seek new markets and resources, towns sprang up, or grew from villages. These towns were trading posts where ethnic groups could bring goods to middlemen, who would in turn sell them to the Swahili traders who plied the caravan routes. Bringing desirable items such as metal tools, weapons, beads, and cloth for the inland African market, Swahili traders brought coastal culture to the interior.

The arrival of coastal goods and Swahili tradesmen helped Swahili culture and language make inroads into the interior of East Africa. By the eighteenth century, Arab-inspired Swahili design could be seen in the architecture, clothing, and food of tribes from the Indian Ocean clear to Uganda. In caravan towns throughout the interior, mosques sprang up to serve the needs of tradesmen and to convert the local populace. Although most tribes did not convert to Islam, some powerful ethnic groups did become Muslim during this period. In the nineteenth century, the *kabaka*, leader of the Buganda people, converted to Islam, largely attracted by the promise of literacy provided by study of the Koran, the Muslim holy book.

Swahili culture was not the only new culture and ethnicity introduced to the interior of East Africa as a result of Arab trade. In the early nineteenth century, Seyyid Sa'id, the Omani Arab leader of Zanzibar, invited financiers, accountants, and clerks from Asia, especially from India, to work in the great trading towns of coastal East Africa. Asians still share a strong presence in East Africa today, maintaining their religion, Hinduism, as another aspect of East African culture.

Some Long-Range Consequences of the Arab Influence in East Africa

The Arab presence in East Africa has influenced the culture, lifestyle, and economy of East Africa in a multitude of ways, some obvious and some more subtle. The official language of Kenya, Tanzania, and Uganda, Kiswahili, is obviously a byproduct of the Arab presence, as is the development of the Swahili ethnic group.

What is not so obvious, however, are the subtle ways Arab trade altered the economy of the region, and the traditional cultures and lifestyles of the East African people. As a result of the international trade spurred by the arrival of Arab sailing ships, East Africa became irrevocably tied to the world market. But many historians believe that this tie to the world market came at a terrible price for East Africa and resulted in the ultimate present-day underdevelopment of the region.

Economists and historians label as "underdeveloped" nations that are wealthy in natural and human resources but have a lower standard of living and are less in-

Asians in East Africa

In the 1800s, workers from southern Asia—mainly India—were invited by the sultan of Zanzibar to come to the East African coast to work in the financial sector. Working as bookkeepers and in the customs house with the import-export trade, South Asians became an essential part of the East African economy. And like all immigrants, they also became a part of the culture of the region, bringing new languages, traditions, and a new religion: Hinduism.

Today South Asians live in small towns and large cities throughout East Africa. They often own small businesses that serve the local populace. Their business acumen, different culture, religion, and language have often been a source of tension in East Africa, however. Leaders jealous of their success have often made them scapegoats of a poor economy, in so many words telling the East African people, "You are poor because the Asians are rich."

Anti-Asian sentiment culminated in 1972 when Ugandan president Idi Amin ordered the expulsion of all noncitizen Asians. All but four thousand Asians left the country—mostly for Britain—leaving their shops, businesses, homes, and wealth to be confiscated by the government. In 1991, in a goodwill gesture calculated to please international human rights critics, another Ugandan president, Yoweri Museveni, invited all expelled Asians to return. Many did, but many more chose to remain in Britain.

Four Kenyan Hindus. South Asian immigrants have experienced harsh opposition in East Africa.

dustrialized than countries characterized as developed. Because Arab trade was largely a process of extraction (taking natural resources without a fair-value exchange), East Africa gained little economically from the relationship. Those who profited most from the trade were not individual tribesmen who were the primary producers and procurers of the goods but the Arab tradesmen and other foreigners. The wealth of East Africa was thereby largely exported to other countries.

The export of a wealth of human resources as slaves, begun with the Arabs and continuing with the Europeans, also had a detrimental impact on East Africa's economy. As a multitude of workers were taken from the hinterlands to be used as slaves first in Arabia and then on the clove plantations of the island of Zanzibar, the ethnic groups of the interior lost valuable man power and brain power that could have helped them advance culturally.

The extraction of wealth from the interior and the export of man power were not the only aspects of Arab trade that historians believe were damaging to the nascent East African economy. Some goods imported into East Africa were also detrimental. With the exception of iron tools, most imports were useless at best; glass beads, mass-produced pottery, and the like could do little damage but provided no particular benefit to the economy. The import of cheap cloth was another matter.

African manufacturers produced cloth for the local market. Imports of inexpensive cloth from Asia, however, damaged this industry by flooding the market with an inexpensive product with which the East African companies could not compete. Because of Arab trade, some historians have argued that the cloth industry did not develop as it might have. Thus, between the extraction of resources, the introduction of cheap cloth, and the importation of items not helpful in the development in the region, East Africa entered its new era of European trade underdeveloped and in a state of flux and change.

Chapter 3

Europeans in East Africa

Like the Arabs, Europeans first came to East Africa as commercial explorers. And, like the Arabs, they ultimately exploited the region for their own economic gain. But European attitudes toward East Africa were quite different from those of the Arabs. Whereas the Arabs intermarried with coastal people, blending cultures and creating family trading dynasties, the Europeans saw themselves from the start as colonizers, always at odds with a colonized people. From the late fifteenth century, when Portuguese explorers came upon the riches of the East African coast, Europeans were gripped by a desire to explore, exploit, and own the wealth of the region.

Three European nations—Portugal, England and Germany—made East Africa the object of their desires. These nations used three methods to gain access to, and ultimately colonize and possess, the region, its people, and its resources: exploration, evangelism, and exploitation.

Explorers covered East Africa from the Indian Ocean to the lakes of the west, mapping out the physical and cultural geography of the region. Most explorers were sponsored by governments or commercial enterprises in their country of origin. Although they may have been personally motivated by passionate curiosity, a sense of adventure, and probably patriotism, they also held out hope of financial gain from their exploration. As agents of European nations, they made alliances with ethnic groups throughout the region, and paved the way for later and concurrent waves of European immigrants—marketeers, politicians, and Christian evangelicals, also known as missionaries.

Missionaries, motivated by a desire to convert East Africans to Christianity and end the East African slave trade, set up missionary stations, churches, and schools in villages throughout East Africa. These stations would later serve as bases for European governments eager to make contacts and claim territory in the region. Missionary stations would also serve as centers of commerce.

European commercial interests, eager to exploit the natural resources of the region, used the local contacts made by missionaries and explorers to make headway into the countryside. Invariably, these commercial ventures were tied up with the colonial interests of European governments. To protect commercial interests, governments made formal alliances with ethnic groups and built infrastructures such as railroads to establish a claim over the region. These claims led to the establishment of formal colonies and protectorates, which ultimately led to the formation of the modern states of East Africa.

Exploration

The first Europeans to reach East Africa were explorers motivated by clear commercial interests. In 1498, Portuguese trading vessels rounded the tip of Africa and entered the Indian Ocean. Searching for a trade route to India that bypassed the Muslim-controlled waters of North Africa and the Mediterranean, the Portuguese were amazed to see the wealth and prosperity of the Swahili city-states. They later returned with warships, and for the next two hundred years, until their expulsion by Omani Arabs, the Portuguese dominated the East African coast, extracting gold, ivory, and slaves.

Seventeenth-century Portuguese colonists in East Africa. The Portuguese were the first Europeans to explore East Africa.

The East African Slave Trade

The European slave trade in East Africa began around 1770, much later than the West African slave trade, which began in the 1500s. And for one hundred years, until the British convinced the sultan of Zanzibar to close the slave market there, the slave trade thrived in the region.

For two centuries, the European slave trade focused on West Africa. Then, in the late 1700s, economic and political conditions caused Europeans to consider East Africa as a source for slaves. The French began developing new sugar plantations on the islands of the Indian Ocean. They looked to Arab and Swahili traders, who had been supplying slaves to the Arab market for some time, as a resource for slave labor.

New sources for slave labor were also needed in South America and the Caribbean Islands. High prices for West African slaves, as well as a spate of antislavery raids on ships traveling from West Africa to South America, made the long journey from East Africa a more attractive prospect. An active European slave trade thrived in East Africa for a hundred years, until the industrial revolution diminished the need for slaves and public opinion began turning to abolition.

While Portugal dominated the coast, however, it did little to explore the interior of East Africa. In fact, until the end of the eighteenth century, Europe had little interest in exploring East Africa or learning anything that would contradict its views of Africa as the "dark continent." When an early explorer, Scottish nobleman James Bruce, returned from an expedition to search for the source of the Nile River in 1768, his reports of the region were ignored. "Bruce," writes historian Kevin Shillington,

> found that few would believe his stories of the splendors of Ethiopia and in particular its royal court at Gondar. . . . Despite centuries of coastal trading contact, Europeans were still remarkably ignorant of Africa, its peoples and their history. European interest in Africa, however, was about to be awoken.[14]

It was not until the end of the slave trade that European interest in exploring East Africa was awoken. As the need for slaves diminished with the introduction of the industrial age, Europeans began to believe that slavery was wrong and that Europe should try to end the trade altogether. Since they were not to make money from slaves, Europeans developed an interest in discovering what else

Africa had to offer. Thus, commercial exploration—the search for raw materials, markets, and already established trade routes—was born. One of the first groups to form for the purpose of commercial exploration was the African Association. Spurred by the hope of wealth and adventure, dozens of explorers sponsored by this British group set out for Africa between 1788 and 1877.

Governments and commercial enterprises sponsoring explorers hoped to lay claim to land and resources or discover new markets for the products of industrialized Europe. For this reason, much of the information brought back by explorers of the era was kept secret so as to benefit only one company or nation. The constitution of the African Association, for example, dictated that the information gathered by its explorers be kept secret so that it benefited only Britain. In 1799, Mungo Park, the African Association's most successful explorer, expressed the purpose of his journeys: "rendering the geography of Africa more familiar to my countrymen, and . . . opening to their ambition and industry new sources of wealth, and new channels of commerce."[15]

Eighteenth-century Scottish explorer James Bruce was one of the first Europeans to venture into Africa's interior.

Wealth and commercial gain were not the only reasons explorers flocked to East Africa. The romantic adventure of exploring the unknown was very much a part of the spirit of the period. This account by a member of the 1886 expedition of German count Samuel Teleki von Szek shows the romantic attitude of many explorers of the time:

> Very arduous has been their [previous explorers'] work, and many are they who have fallen victims by the way; but others, imbued with a similar zeal for the furtherance of scientific knowledge, have ever been ready to take their places and to follow the rugged paths leading to the heart of the great continent. And no wonder! For mighty is ever the fascination exercised by the unknown, and, to the enthusiastic spirit, no charm can excel that of devoting every power to a noble aim.[16]

Mungo Park wanted to explore Africa to gain "new sources of wealth, and new channels of commerce."

Evangelism and Its Fruits

Just as the exploration of East Africa began as a mixture of noble aims and commercial expediency, so too were the ideals of the evangelical movement mixed up with commerce and politics. Several trends, both religious and political, combined in the 1800s to create a wave of enthusiasm for evangelism, or the drive to gain converts to Christianity.

The first trend to spur the missionary movement to Africa was a European Christian revival, a renewal of Christian worship and beliefs. Preachers of the European revival proposed the belief that it was the duty of every Christian to spread the message of Christianity throughout the nonbelieving world. In the nineteenth century, because of European ignorance of Africa, many Europeans believed that most Africans had no spiritual beliefs at all. While nothing could have been further from the truth—East Africans had rich and varied spiritual lives—this mistaken understanding fueled the evangelical fire.

The second trend to fuel the missionary movement in East Africa was the drive to end the slave trade. Horrible stories of the Arab slave trade circulated throughout Europe. As a result, abolitionist organizations, organizations seeking to end slavery in East Africa and beyond, abounded in Europe. Many of these were connected with Christian evangelical churches. Throughout East Africa, these organizations set up missions to free slaves and convert Africans to Christianity. Many set up commercial enterprises to provide work for freed slaves and income for the missionaries. Thus, the connection between evangelism and commerce was established.

The commercial work of missionary societies, combined with the lack of any other political power in the hinterland of East Africa, gave missionaries extraordinary regional powers. Missionary stations became not only centers of commerce but also centers of political and cultural influence. Europeans—both government dignitaries and leaders of commerce—made the missionary stations their base when visiting East Africa.

Dr. Livingstone, I Presume?

The man largely responsible for focusing the European public's attention on the East African slave trade was Dr. David Livingstone, a physician and missionary from England. Livingstone had traveled twice across the interior of Eastern and Central Africa by 1856, exploring rivers and trade routes and documenting the horrors of the slave trade. A confirmed believer in the powers of both Christianity and commerce, he proclaimed that commerce and Christianity, riding together into East Africa on the rivers of the region, had the power to end the slave trade and save the souls of the African people.

Livingstone is perhaps best known today not for his views or deeds but for his disappearance. When engaged in missionary exploration in the 1860s, Livingstone lost contact with his sponsors and was presumed to be dead. Henry M. Stanley, one of the most famous explorers of his time, was sent to search for him. When he finally found Livingstone, living in ill health in a Central African village, Stanley greeted him with the famous words, "Dr. Livingstone, I presume?"

Livingstone's greatest gift to the evangelical spirit of his times was his inspirational views. His death in the early 1870s ignited the evangelical movement with an enthusiasm that continued till the 1900s. Livingstone's evangelical vision of combined commercial and missionary work took form in many missionary stations throughout East Africa. Most missionaries focused on commerce as an offshoot of antislavery or other work. Some established colonies of freed slaves along trading routes. Others hoped to use commerce to break the slave trade. The Livingstonia Central African Trading Company, for example, the work of Scottish missionaries, hoped to establish an ivory commerce that would undercut the Arab ivory trade without using slave labor.

Physician, missionary, and explorer David Livingstone strove to end the slave trade in East Africa.

Missionary stations also became the primary means for introducing European culture and establishing political connections to local ethnic groups. Mission schools, using European educational methods and their own languages, taught basic literacy to East Africans and exposed them to European values. It was from these schools that an educated elite would emerge in the twentieth century, eager to take over the political reins of developing East African nations. The political power of missionary stations and their relationships with local ethnic groups became most important after 1884 as European nations began to scramble for formal colonies in East Africa.

A 1905 photo shows a Christian missionary surrounded by a group of novice brothers in Uganda.

Exploitation

The European "scramble" for East African territories is closely tied to each nation's commercial interests. Explorers and missionaries may have been the vanguard in East Africa, coming first and introducing the local ethnic groups to European people and ways, but European commercial and political entities were the main forces of colonization. They used these earlier connections in the scramble for influence and power called "free trade."

Under the informal system of free trade, any nation was free to use commercial, political or religious means to create spheres of influence—areas of East Africa where tribes were loyal to particular European nations. Europeans used free trade in East Africa to gain influence over the region and gain access to raw materials and markets without assuming the responsibility of a formal colony.

Like all commercial activities, the movement to establish free trade in Africa was born of self-interest, not the interests of those who happened to live in the target area. In East Africa, free trade focused on finding resources to exploit other than slaves, and markets in which to sell European goods. Commerce that did not include slaves, called "legitimate commerce," was not much better for the local economies than slavery had been. Legitimate commerce included the growth of agricultural produce for export rather than for the nutrition of the local people, the large-scale extraction of animal products such as elephant tusks, and the exploitation of other natural resources. In all, this sort of exploitive commerce hurt the economies of East Africa in ways that have lasted up to the present time. Historian Kevin Shillington explains:

> The establishment of "legitimate commerce" did not long allow African states to develop their own economic strength and independence. In the first place those that benefited from the trade were a small minority of wealthy rulers and merchants. There was little improvement in the social and economic well-being of the bulk of the population. Indeed, many saw their living conditions and levels of personal freedom decline as their labor was harnessed to increased production and transport for the export trade. Secondly, the principal imports from Europe—cloth, alcohol and firearms—did nothing to strengthen indigenous African economies. Finally, those states that did develop their export trade soon found their independence threatened by direct interference from their European trading partners.[17]

The nineteenth-century trading partnerships between East African ethnic groups and European nations took on different forms for Britain and Germany. These two countries were now the principal European nations in East Africa since Portugal had long ago been expelled by the Omani Arabs. Britain enjoyed secure commercial and political control over its sphere of influence by means of influencing the decisions of Seyyid Sa'id, the Omani sultan of Zanzibar, and maintaining commercial agreements.

In 1884, however, the British government's exclusive claim on the territory was challenged by a powerful German presence in Tanganyika, what is now mainland Tanzania. By making alliances and signing treaties with powerful tribes in Tanganyika, Germany had established a commercial and political presence in the region. The scramble for East Africa was on.

The scramble for Africa was formalized in 1884 at the Berlin Conference, a meeting of all European nations with interests in Africa. The Berlin Conference effectively divided up Africa among the European nations by proclaiming that "a European claim to any part of Africa would only be recognized by other European governments if it was 'effectively occupied' by that particular European power."[18] Within

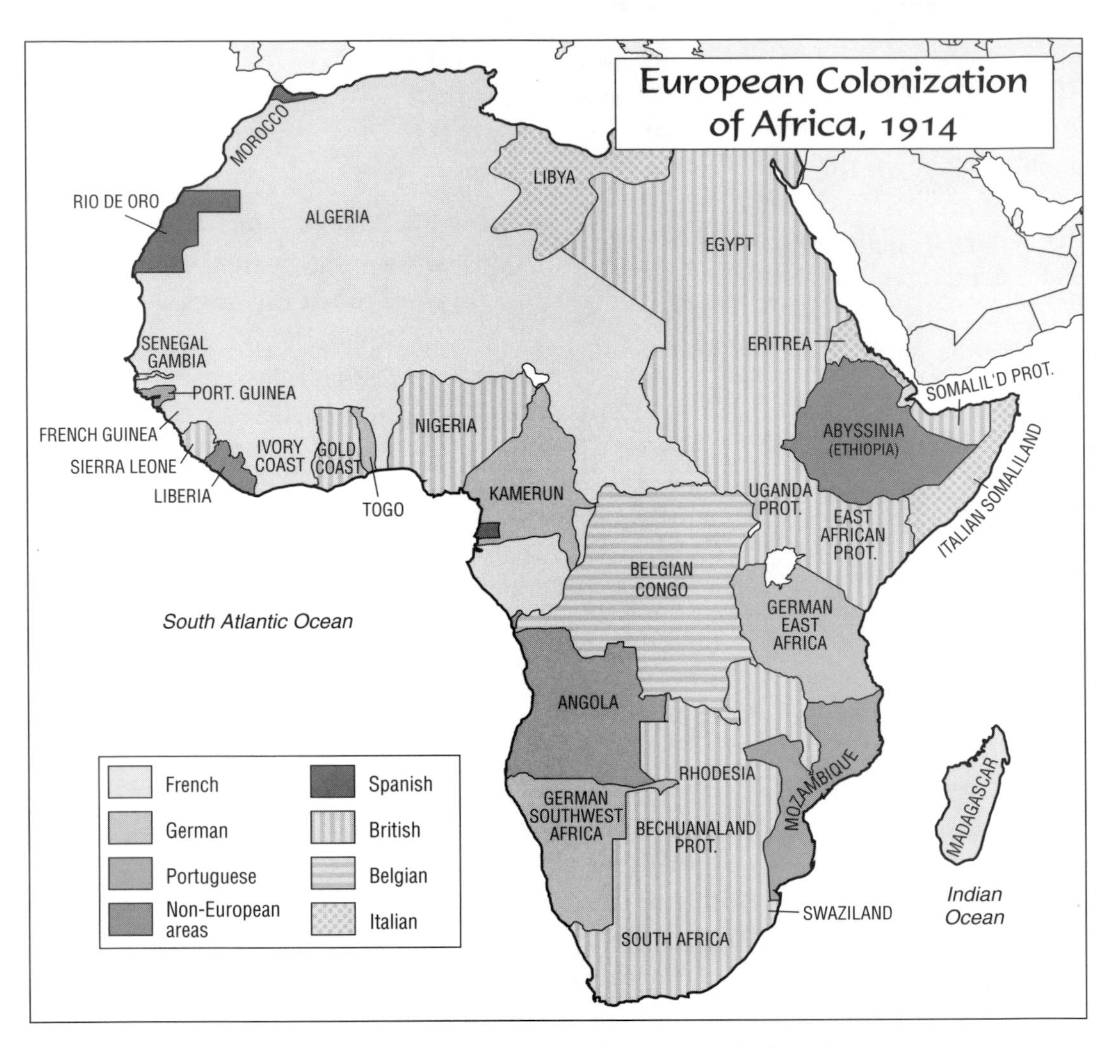

days of the conference, Germany claimed a good portion of East Africa, calling it German East Africa and later Tanganyika. Britain gained the Uganda Protectorate in what is now Uganda and the East African Protectorate, renamed Kenya in 1920. Portions of the coastal region remained in the hands of Seyyid Sa'id's heirs, the Omani Arabs, and their Swahili descendants until British pressures pulled it slowly into British hands.

Colonial Rule

With portions of East Africa now formally in their hands, Germany and Britain expected their new colonies to both pay for themselves and stay under control. To ensure these results, the two nations systematically exploited the region for commercial gain and firmly managed the various ethnic groups. Laws forced the Africans to provide land, products, or revenue to British and German companies and governments; these laws also helped the colonizers maintain control over local populations.

European farmers and concessionaires—companies to whom land was leased—were encouraged to come to East Africa to farm, develop, and exploit natural resources. These companies and individuals had the full legal backing of Britain and Germany but little sense of responsibility to the local people. Local African groups were encouraged—and often forced—to grow crops for export or collect raw materials for concessionaires and farmers. The force used to compel such labor was unethical, violent, and even deadly.

The land given to concessionaires and farmers was traditional tribal land. Many ethnic groups—such as the Nandi and Kikuyu in Kenya—were forced off fertile and productive lands and transported onto poor reserves much like the reservations to which Native Americans were "removed" in the nineteenth century. Tribe members who refused to leave their land were called squatters and were subject to

Colonizing nations often used the local people as labor. Here, workers build a railway from Dar es Salaam to Morogoro in German East Africa in about 1920.

arrest and imprisonment. Others were permitted to live on their traditional lands only if the white "landowners" needed their labor.

To control the supply, cost, and movement of labor, pass laws were developed. Pass laws made it mandatory for East African men to carry identity cards called *kipande* at all times; these cards stated who their employer was and how much they were paid. Breaking these laws carried heavy consequences: It became a criminal offense to refuse to work or to be in the wrong place at the wrong time.

It also became a criminal offense to refuse to turn over a portion of one's produce or income as taxes to the colonial government. Taxation was yet another means used to compel East Africans to work for Europeans. The governor of Kenya spoke of taxation in the *East African Standard* newspaper in 1913:

> We consider that taxation is the only possible method of compelling the native to leave his reserve for the purpose of seeking work. Only in this way can the cost of living be increased for the native. . . . It is on this that the supply of labour and the price of labour depend.[19]

In addition to establishing taxes to compel East Africans to work for wages, the British and Germans also controlled the amount of money ethnic groups could earn from their products. They set up official export boards—associations controlling the sale of produce abroad—throughout the region. British and German middlemen from these boards bought African produce at prices far below the fair world-market value. Then the Europeans resold the African produce at a higher price on the world market, reaping a tidy profit and leaving the East African people in poverty.

Unhappy with their lot of poverty, taxation, low wages, loss of land, and violent repression, East African ethnic groups began to demand independence. Independence from Germany was achieved in the early part of the twentieth century. Germany lost its East African territories after World War I, and Tanganyika was turned into a League of Nations trust, and later a United Nations trust territory under British rule. By the 1920s, Britain controlled all of East Africa. At last, the stage was being set for the creation of independent nations.

Resistance, Rebellion and Independence Movements

The movement toward the formation of independent East African nations can be said to have begun with the small, ethnically based rebellions, resistance movements, and organizations that sprang up throughout the British and German tenure in East Africa. For much of this time, Britain and Germany managed to keep rebellion at bay by fanning the flames of ethnic tensions in the region. By playing one tribe against another—for example, giving one tribe the land of another—the Europeans could keep the Africans focused on intertribal problems rather than on the larger regional conflict against the Europeans.

Maji Maji and Mau Mau Rebellions

Two major rebellions rocked East Africa during the colonial period: Maji Maji and Mau Mau. Maji Maji was the first major violent rebellion in colonial Tanganyika. It was unique in that it spread spontaneously across ethnic lines yet had no leaders. The rebellion spread through southern Tanganyika as a response to the violence and intimidation used to force Tanganyikans to pay "head taxes," taxes simply for being alive, and to work on German plantations. Hundreds of foreigners were attacked—European, Swahili, and Arab missionaries, administrators, and clerks. Historian Kevin Shillington in *History of Africa* describes Maji Maji in this way:

"It was a deliberate attempt to overcome the problems which had crippled earlier African resisters to European conquest, namely, lack of African unity and the European machine-gun. [Rebels] sprinkled their bodies with magic water known as *maji-maji* which [they believed] would turn the bullets of their enemies into water. It was a simple device which brought the people together."

Mau Mau was an equally violent Kikuyu resistance movement that took place in Kenya in the 1950s and tapped into the Kikuyu's spiritual ties to their land. While scholars disagree about the source of the word *Mau Mau*, many Kikuyu believed it to be a derogatory term suggesting greed or childishness coined by enemies of the movement. Mau Mau fighters fought from the forests of the highlands, taking part in tradition-based rituals, reciting powerful oaths and vowing to kill those who had taken their land. In *Mau Mau Memoirs: History, Memory, and Politics*, Marshall S. Clough writes,

"These vows commit [the initiate] to struggle for freedom, never to forsake or sell the land to Europeans, to help the secret movement with firearms, money, or anything else needed, to obey superiors, and never to betray the movement to its enemies. For example, J. M. Kariuki and his comrades in the Rift Valley swore:

I speak the truth and vow before God

And before this movement

The movement of Unity

The Unity which is put to the test

The Unity that is mocked with the name of "Mau Mau,"

That I shall go forward to fight for the land,

The lands of Kirinyaga that we cultivated

The lands which were taken by the Europeans

And if I fail to do this

May this oath kill me."

But ethnic tensions could not keep East Africans from unifying forever. Within and between ethnic groups, in both rural and urban areas, young missionary-educated elites began forming nonviolent political "associations" such as improvement associations and trade unions. These were formed mainly to fight for fair wages and to educate their members. Youth associations were often formed in rural areas and dealt with local issues of land reform and education. All the associations played a role in educating East Africans and unifying them under nationalist ideals.

Two Mau Mau fighters. The move toward independence for East African countries began with rebellions such as the Mau Mau.

Jomo Kenyatta is led into a courthouse in Kenya in 1952, charged with leading the Mau Mau Rebellion against the British.

Though local associations were useful in encouraging East African ethnic groups to come together in a new way, it was large-scale national political movements that finally succeeded in drawing East Africans together to press for independence. The leaders of these nationalist movements—Jomo Kenyatta in Kenya, Julius Nyerere in Tanzania, and Milton Obote in Uganda—worked to channel East Africans' frustration with colonial rule into national movements that would cross ethnic lines and bring their nations together. And each man would go on to become the first leader of his new nation when independence was finally achieved in the early 1960s.

The Mixed Outcome of Colonization

From their very first contact with East Africa until independence was finally achieved, Europeans saw East Africa as a source of wealth and promise. Explorers, evangelists, entrepreneurs, and politicians all sought to mine the commercial and human resources of the region, looking for converts to both Christianity and the free market economy. The outcome of this exploration and exploitation benefited Europe greatly. Resources—human, natural, and economic—were extracted from the region for Europe's gain. East Africa, however, was left to struggle to maintain a healthy standard of living for its people. Today, the modern nations of East Africa continue their struggle for economic development and political independence from their former colonial powers.

Chapter 4

Cultures of East African Ethnic Groups

The cultures of the people of East Africa are the traditions and beliefs that guide them from the beginning of their lives to the end. Cultural traditions connect individuals to a larger sense of themselves and their community and ensure that there is continuity in an ethnic group that will help them survive as a people. Jomo Kenyatta, the first president of Kenya, once said this about the importance of cultural traditions: "It is the culture which he inherits that gives a man his human dignity."[20]

While cultural traditions root ethnic groups in their past and give them a sense of themselves as persons of dignity, culture is also changeable. Ethnic cultures change as tribes interact with each other and with the outside world. Where tribes come together and mix, cultures change.

Many East African ethnic groups share cultural features. Often, unique cultural rituals are shared, but the tribes' lifestyles—how they make a living on the land—are completely different. The Maasai and Kikuyu tribes of Kenya, for example, have been neighbors for hundreds of years and share many rituals, such as rites of passage. Their lifestyles, however, are completely different: The Kikuyu are agriculturalists while the Maasai are pastoralists.

Cultural features such as rites of passage, the roles people play in their communities, and the rules and values that guide them distinguish the various cultures of East African ethnic groups. No two groups are bound by the exact same set of cultural principles and traditions, and it is these principles and traditions that make East Africa a colorful and vibrant region.

Roles

In many East African cultures, the roles people play in a community, tribe, or clan are often determined by binding customs that consider age, sex, and tradition. Women and men, boys and girls, old people and young have specific jobs in the community assigned to them that help the tribe function and thrive.

In this village in Kenya, women raise goats for meat, milk, and skins. Gender, age, and tradition often dictate which jobs people perform.

Among the Gusii, an agricultural ethnic group who live east of Lake Victoria Nyanza in Kenya, young girls frequently play the role of *omoreri*, or caregiver. According to the authors of *Childcare and Culture: Lessons from Africa*, an *omoreri* is a female relative six to eleven years old who cares for an older infant or toddler

> for periods during the day when the mother is busy with other tasks. . . . [Her role is] to protect the baby from harm, largely by holding, and to respond rapidly to distress with soothing and feeding. . . . They are not, for the most part, instructed to play with the baby, although they are free to do so according to Gusii kinship norms, and mothers *expect* that the *omoreri* will play with the baby.[21]

Although there are many similarities between the jobs of American baby-sitters and Gusii *omoreri*, the two roles are quite different. American baby-sitters are usually teenagers or older preteens who are paid to look after children to whom they are not related. An *omoreri*, on the other hand, is a young girl who performs an unpaid role for extended family members simply because it is expected of her by her culture.

Girls are not the only members of ethnic groups who are bound by strict roles. Boys, too, play roles that are determined less by their abilities and interests than by their age. Among the Maasai and Samburu tribes, for example, young men age fifteen to twenty-five have a very important role to play in the community as *moran*, or warriors. *Moran* have a vital practical as well as symbolic function as protectors of the tribe. Former Maasai warrior Tepilit Ole Saitoti provides an English-language version of a *moran* song:

Young Maasai men in Kenya perform a dance. Maasai men age fifteen to twenty-five are the society's warriors, or moran.

"Young are the warriors, and we feed them the best of our meat. Healthy, they will protect our herds from enemies and famine. And they will stop all the foes of our people from encroaching upon us."

These words, sung by senior warriors to their younger comrades during their feasting camps, express something of the significance and the almost magical powers attributed to the title *ilmoran*. . . . Whenever there is work requiring strength or courage people will ask, "Are there no warriors around today?" Warriors are needed both for simple tasks—capturing a cow to be slaughtered for a ceremony and decorating it with bells, wrestling with one to be branded—and for dangerous ones—protecting the herd from lions or subduing a crazed rhino charging through the kraal [homestead]. Maasai warriors in their prime seldom fall short in the performance of their duties. When praised, they will modestly answer, "All we did is what Maasai warriors are supposed to do."

I remember from my own experience as a warrior how . . . we were totally trusted by our community for protection, and we tried to live up to their expectations.[22]

Individuals are trusted by their communities to play the roles they have been assigned. Often, the health, safety, and prosperity of their family and tribe depend on their adherence to traditional roles. But the roles an individual plays change throughout his or her lifetime. For this reason, the transitions from one role to another, from one age group to another, and the rituals celebrating these transitions are an important aspect of the cultures of many East African ethnic groups. These rituals are called rites of passage.

Rites of Passage

Rites of passage are rituals undertaken by a tribe to mark an individual's or age group's transition from one stage of life to the next. Rites of passage carry individuals from birth to death, tying them to their community and strengthening the interdependence of individuals and their tribe. These rituals also ensure that individuals respect and follow the most important rules set forth by the culture.

Rites of passage celebrate birth, the end of infancy (usually marked by weaning between ages two and three) the passage from childhood to adolescence, marriage, elderhood, and death, which many tribes consider to be the beginning of the afterlife. Rites of passage can include both pain and celebration. Some rites involve dancing, singing, and feasting, while others include dangerous and painful surgical procedures. Often, both are involved.

Rites of passage vary from ethnic group to ethnic group, though many aspects are shared between tribes who have lived near each other at different points in history. Many ethnic groups, for example, perform rites of passage for individuals alone,when they reach a certain stage in life—such as when a girl begins to menstruate. Other tribes, however, perform rituals for groups of people called age sets. An age set is a group of people of similar age who go through a rite of passage or initiation together and, for the next ten or fifteen years, contribute to the community by fulfilling the role into which they have been initiated. This tradition is shared by many ethnic groups, both pastoral and agricultural. An example of this age-set tradition is the Maasai *moran*, who are initiated together and work as warriors for ten or fifteen years until they go through another rite of passage marking their transition to elderhood.

The rite of passage marking the transition from childhood to adulthood is common to most ethnic groups in East Africa. For many tribes, it involves the most dangerous and dramatic of initiation rituals: circumcision, the ritual removal of parts of the genitalia. Female circumcision has become controversial throughout East Africa, but in many tribes—such as the Samburu of central Kenya—this rite of passage is a deeply entrenched tradition thought to be an essential part of becoming both a woman and a fully invested tribe member.

Samburu Rites of Passage

The Samburu are Paranilotic pastoralists who herd cows, goats, and some camels in

Young Samburu men and women dance during circumcision celebrations in Kenya. Circumcision is part of the Samburu rite of passage from childhood to adulthood.

arid northern-central Kenya. They are related to the Maasai tribe and practice many similar rites of passage. At about age fifteen, boys who aspire to become junior *moran* go through elaborate rituals in which they must show courage in the face of fear and pain. After singing a brave song and uttering boasts welcoming their

Female Circumcision Controversy

Circumcision of males—the removal of the foreskin of the male genitals—is common in many cultures throughout the world. It is done either shortly after birth, as in many developed nations and among Jewish people worldwide, or at adolescence, as in many East African ethnic groups.

Female circumcision is much more rare. It is virtually unheard of in the developed world and is becoming increasingly controversial elsewhere. Many East African ethnic groups view the ritual removal of parts of the female genitalia as an essential part of becoming a woman. But many people throughout the world believe the procedures to be cruel and dangerous. According to the human rights organization Amnesty International, an estimated 2 million girls worldwide undergo circumcision every year. Fifty percent of the women in Kenya, 10 percent in Tanzania, and 5 percent in Uganda have submitted to some form of this practice. The website of Amnesty International provides a gruesome description:

"Sometimes a trained midwife will be available to give a local anaesthetic. In some cultures, girls will be told to sit beforehand in cold water, to numb the area and reduce the likelihood of bleeding. Most commonly . . . no steps are taken to reduce the pain. The girl is immobilized, held, usually by older women. . . . Mutilation may be carried out using broken glass, a tin lid, scissors, a razor blade or some other cutting instrument. . . .Thorns or stitches may be used to hold the [injured area] together, and the legs may be bound together for up to 40 days. Antiseptic powder may be applied, or more usually, pastes—containing herbs, milk, eggs, ashes or dung—which are believed to facilitate healing."

Female circumcision has become a hot-button issue worldwide as more people migrate from developing nations to industrialized countries. As the practice has come to the attention of health care workers in the developed world, protests against it have increased. In 1979, at a conference in Khartoum, Sudan, the United Nations World Health Organization "condemned the mutilations as disastrous to women's health and . . . indefensible on medical as well as humane grounds." Many tribes continue the practice, however, defending it as an essential cultural rite of passage.

The Samburu are expected to bear the circumcision procedure without flinching. Pictured here are Samburu boys in Kenya about to be circumcised.

circumcision, young men are led to a special place. Circumcision is conducted by elders without anesthetic, and under conditions that are primitive by Western standards. Initiates maintain their honor, and the honor of their families, by not flinching or moving a single muscle during the excruciating four-minute procedure. To move at all during initiation is to bring dishonor upon subsequent generations, as well as one's elders.

After undergoing the ritual circumcision, boys still do not bear the honor of being *moran*. A month later, following an initial period of the food and lifestyle restrictions common to *moran*, the *ilmugit* of the arrows ceremony takes place. Anthropologist Paul Spencer, in his book *The Samburu*, describes the ceremony:

> The *Ilmugit of the Arrows* is the first of [several] *ilmugit* ceremonies, and in the course of its performance all the initiates become junior *moran*. Each initiate has two ritual partners who slaughter his *ilmugit* ox or goat on his behalf and enter into a relationship with him . . . of brotherhood. . . . [They] lead him to his mother's hut where he vows that he

> will no longer eat meat seen by any married woman: such meat is *menong'*, despised food. . . . The avoidance of this food is the most characteristic restriction observed by Samburu *moran*, and they themselves see it as a determining criterion of *moran*hood. . . . A youth who does not observe this restriction is something less than a *moran*—he is behaving like a child. . . . Having made his vow to his mother, the initiate is now a *moran* and he can put red ochre [a red mineral dye] on his head and body for the first time in his life as a decoration.[23]

Throughout their service as *moran*, youths will perform other *ilmugit* ceremonies marking their increasing maturity. Finally, at about age thirty-one, the last *ilmugit* ceremony, the *ilmugit* of the milk and leaves, takes place. After this ceremony, each man and the woman he has recently married are blessed by the elders, and the former *moran* are finally free to give up the food restrictions imposed on all *moran*. They have become elders.

Becoming an elder does not mean that a man has reached the pinnacle of the Samburu community, however, or that his learning and growth are complete. It simply means that his role has changed from warrior to family man and decision maker. Elders play a special role in the Samburu community as decision makers, guides, and leaders of their families and clans. And as they age, elders continue to go through a variety of rites of passage marking their continuing transitions and increasing maturity.

The initial rite of passage for Samburu girls is similar to that of boys in its requirement that girls show courage and endurance during an excruciatingly painful circumcision surgery. For girls, however, initiation does not end in a transition period between childhood and adulthood—*moranhood*—as it does for boys. Girls pass from childhood to adulthood in one step: marriage.

Marriage is the culmination of a girl's initiation period. During the period leading up to her initiation ceremony and subsequent marriage, Samburu girls are taught the things they will need to know as a woman. Shortly after her circumcision is complete, at around age fourteen, a girl is married to an elder, usually age thirty or older, from outside her clan and is sent to live far from her own family. She is suddenly responsible for building and maintaining her own house and household and caring for her own herd of cattle, given as a marriage gift called bridewealth. For Samburu girls, puberty marks their introduction to adulthood.

In many Bantu-speaking tribes, girls and boys make the transition to puberty as individuals. A girl's rite of passage is conducted upon beginning menstruation. Among the Zaramo people of Tanzania, for example, a cycle of female rites begins shortly after a girl's first menstrual period and continues through the birth of her first baby until the maturity rites of that baby as it passes to childhood. In her initial rite, according to anthropologist Marja-Liisa Swantz,

> The girl is separated from certain categories of people, her seclusion is well guarded. . . . [Then] the female representatives of the husband's [the girl's father's] family perform the rites with the girl's assistant and the girl herself. The coming out dancing and drumming are marked by wide neighbourhood participation as a sign of great rejoicing. . . . Should the girl not yet have. . . a husband, naturally this day then publicly announces her eligibility for marriage.[24]

Marriage

Marriage is a rite of passage in many cultures around the world that marks an individual's transition from the life of a child to that of an adult. East Africa is no exception, and the types of marriage ceremonies that take place there are almost as varied as the tribes themselves. In some tribes today, Christian and Muslim marriage rites have replaced traditional marriage ceremonies. But in many ethnic groups, marriage traditions have remained the same for hundreds of years.

For many tribes, marriages are arranged between families or by community elders, ensuring that not only do the individuals make a good match but also their families are well suited to each another. Some ethnic groups, such as the Samburu and Maasai, are exogamous, requiring members to marry outside their clan. This promotes the connections between clans in the larger tribe and prevents genetic defects that can occur when close relatives marry and have children.

Birth and Death

Marriage, quite naturally, leads to the next life-cycle transition and rite of passage that generally takes place: the birth of a first child. The birth of a child and death of an elder are especially important life-cycle transitions. Special rituals associated with birth and death assure that a person belongs to his tribe from his very first breath to his very last.

Among the Gusii, agriculturalists of western Kenya, a woman's first childbirth is attended by as many as thirty elder women. These women do not offer encouragement during labor but ridicule and tease the woman as she delivers her baby. Once the baby is born, however, special steps are taken to protect both mother and child. The placenta, or afterbirth, is buried in the ground to ensure the spiritual safety of the child, and the mother and baby remain in seclusion for three to six weeks. After that time, the *ekairokio*, or ceremonial emergence, can take place. The authors of *Childcare and Culture: Lessons from Africa* offer this description:

> It takes most of a day and involves a sacrifice to the ancestors, libations

of millet-beer, and administration of protective medicines to mother and infants before their emergence. The mother must also hold up the babies to the sun, asking for protection, on the morning of the ceremony. A grass ring must encircle the house before the actual emergence. Every one of these elements is designed to strengthen mother and newborns physically as well as neutralize the interpersonal dangers that might threaten them when they are no longer in seclusion.[25]

In addition to the birth rites involved in welcoming a new baby into the tribe, each culture also has a unique way of dealing

Gikuyu Bridewealth Traditions

Bridewealth is a gift given to a girl and her family when an agreement is reached to allow the girl to marry a suitor. In many tribes, bridewealth consists of livestock. Many ethnic groups consider bridewealth an essential requirement before marriage. In some ways, the rendering of bridewealth to various family members is symbolic of the new relationship being forged between two families. In a practical sense, it assures a girl's family that her prospective husband has the resources to care for her and the children she may someday have. Similarly, many Americans might view the gift of an engagement ring as an essential gesture of a man's symbolic and financial commitment to his fiancée.

Jomo Kenyatta wrote about the Gikuyu (also called Kikuyu) tribe's tradition of bridewealth in his book *Facing Mount Kenya*. In it, the first president of Kenya recounts that if a girl has accepted a boy to be her future husband, bridewealth must then be gathered by the boy's family.

"When the boy's parents return home they begin to collect sheep and goats, or cattle if they are rich, for the first instalment of the dowry, *roracio*; these would be taken by the lover to the girl's homestead and led to the hut of the girl's mother. . . .This instalment is followed by another in a few days, and so on until the number of animals amounts to about thirty or forty. . . .The amount varies from one clan to another . . . although the amount required by the Gikuyu law is thirty sheep and goats."

It is thought that the payment of the bridewealth, or dowry, in installments, rather than in one payment, gives the two families time to get to know each other. Bridewealth can also be like an insurance policy for a new wife: If she is widowed, her new herd will support her in times of need.

A Maasai woman with a baby in Tanzania. For many East Africans, the birth of a child is an important event marked by special rituals.

with death. Funerals and ceremonies to bury or otherwise dispose of dead bodies are common in every culture. Even among groups that have embraced world religions such as Islam or Christianity, the rituals surrounding death maintain traditional African elements. Among the Zaramo ethnic group, for instance, ceremonies marking a death incorporate both Muslim and traditional Zaramo rites. Swantz provides some examples:

> It is not possible to determine which part is of foreign and which of Bantu origin. The custom of burial itself is not very old. According to the tradition of the Zaramo in Kutu, the people were formerly buried by wrapping the corpse in grass and placing it into

Death is viewed in many East African societies as a transition into the spirit world. Here, a mother and children mourn the loss of a loved one in Tanzania.

what was referred to as a rubbish heap, *jalalani*. . . . The actual burying took place originally only when a chief or another respected elder died. Still an earlier custom was burying such a person in a termite hole in the ground. . . . [Today, like in the 1800s,] the burial posture and shape of the grave followed the coastal Swahili custom, but the dead person was made to face the direction of clan origin usually West or South instead of East.[26]

For many cultures, death is simply another life-cycle transition—the transition from being an elder to being an influential ancestral spirit. The opinions of the spirits of ancestors play an important role in guiding the decisions and maintaining the traditions of many ethnic groups, and many tribes perform rituals on an ongoing basis to honor their ancestors. Pat Caplan, an anthropologist based in a Swahili village in Tanzania, elaborates:

The holding of the main funeral feast does not mean that there are no further rituals for those who have died. They become ancestors and as such are entitled to be remembered in various ways, for example in the ritual of ancestral remembrance (*kuarehemu wazee*) held before a circumcision, and also during the seventh month of the lunar calendar when graves are swept, and people hold Koranic readings for their ancestors.[27]

A Maasai's First School Experience

For pastoralists, school is not simply a place to gain academic knowledge. It represents an entire—and often permanent—change of lifestyle and culture, from living a seminomadic life to living in one place. Often, the children of pastoralists have been separated from their families and sent to live in boarding schools.

Tepilit Ole Saitoti recounts his first experience of school and the changes it made in his life in the book *The Worlds of a Maasai Warrior*:

"When we arrived at the school, we . . . were introduced to the man who was to be our teacher. He gave us some food called rice, which looked as revolting as tapeworms, so I refused to eat it. . . .Until I went to school my staple foods had been milk, meat, and in dry seasons, maize. Now for the first time I would taste sweets and biscuits, European and African fruits such as bananas and oranges, and even grains like beans. The weird smell of soap, which stunk at first, would eventually become acceptable.

My life started changing from that day when the two elders entrusted us to the teacher. One of the old men, Kawanara . . . took a book from the table and showed me some scratches, but I could not make sense of them. When he told me it was a picture of man I thought he was joking. I was used to seeing people standing or walking but not represented on a piece of paper. After three months of schooling, so many things became clear. It was as if a film had been peeled from my eyes. I even understood what Kawanara had tried to show me that first day."

After some time at school, Saitoti was sent home for a vacation. Because there was no written version of Kimaa, the language of the Maasai, other Maasai viewed his ability to read and write as magic. Even the simplest display of literacy by Saitoti and a school friend astonished Saitoti's brothers: "Their eyes would practically pop out of their heads and they would say in amazement, 'Gentlemen, gentlemen, these boys! They are prophets, they are prophets!'"

Honoring the dead through religious readings is one way that East African ethnic groups connect with the past and ensure the continuity of their cultures. Through the advisement of elders and religious leaders who are believed to be able to interpret the ancestors' wishes, many ethnic groups maintain the integrity of their cultural traditions by continuing to perform the same rituals their tribe has performed for hundreds or even thousands of years. Despite the respect accorded ancestors, however, the intrusion of the modern world is bringing many changes to the most basic aspects of culture in East Africa.

Cultural Change

Cultures are dynamic, and change is a part of the creation of culture. The changes in East Africa today are not simply the result of interaction among ethnic groups. Education and literacy, brought by government-sponsored schools and religious organizations, and increased access to worldwide media have also changed cultures. As a result, traditional teachings and skills are lost as tools for the information age are gained. As the women and children of nomadic pastoralist tribes settle near towns to attend school and have access to government services, families split up and rites of passage are delayed. Among agricultural tribes, new economic opportunities have drawn young people to the cities, away from their farming villages and traditional way of life.

Ancient Traditions Live On

Despite these modern influences, however, ancient cultural traditions live on. These traditions are physical manifestations of a tribe's beliefs, hopes, and values. And while they are not immune to the influences of the modern world brought about by schools, governments, and new religious beliefs, by continuing these traditions, East African tribes are able to face the challenges of a changing world from the strong base of a stable culture.

Chapter 5

Religion and Spirituality in East Africa

With more than two hundred ethnic groups living in East Africa, it is no surprise that East Africa is home to a great deal of religious and spiritual diversity. The year 2000 volume of an annual survey of sub-Saharan Africa estimated that 50 to 60 percent of the populations of Tanzania and Uganda are Christian. The number of Muslims in the region ranges from 98 percent of the population of Zanzibar to less than 5 percent elsewhere in East Africa. In Kenya, it is believed that the majority of people follow a variety of indigenous, or local, religions. A very small percentage of East Africans—the Asian minority—are Hindu. Regardless of the religion professed by individual East Africans, however, many continue to be influenced in one way or another by traditional spiritual beliefs and practices. Because of this undercurrent of African spiritual traditions, many religions in East Africa show a remarkable number of similarities in rituals of worship, prayer, and celebration; in their reliance on spiritual leaders; and in their expressions of spiritual beliefs and values.

Traditions in East African Prayers, Rituals, and Celebrations

Prayers, rituals, and celebrations are a public expression of spirituality worldwide, and East Africa is no exception. Every religious group has a calendar of regularly scheduled religious observation as well as an openness to ongoing and spontaneous expressions of belief, thanks, and requests for divine intervention. These verbal, visible, and often colorful expressions of spirituality say a lot about the religious aspect of culture in East Africa.

A large part of every religion in East Africa is the expression of thanks to a divine being or beings. This may be an expression of gratitude for the bounty of nature in a good harvest, for plentiful—but not too plentiful—rain, or for abundant births of livestock or children. These expressions of thanks are similar among

A priest walks out of church with his parishioners after a mass at a Catholic mission in Kenya. Christianity is one of the main religions of East Africa.

many ethnic groups in East Africa even when a tribe has become predominantly Muslim or Christian. For example, although Christians kneel and offer prayers of thanksgiving to God, and Muslims prostrate themselves and pray to Allah, both groups may decorate their homes with plants and fruits of the harvest to show the cause of their celebration.

The Samburu pastoralists also show their gratitude with prayers and decorations of plants, giving thanks to *Ngai*, their word for God, for bountiful rains that bring the grass to feed their cattle. When a heavy rain has subsided in the Samburu homeland, every mud-dung hut is decorated with a sprig of green plants placed above the doorway as an offering.

Decorating with plants is a way of offering a visual prayer of gratitude to a divine being. Singing and oratory, or expressive speech, are also common methods of prayer among many East African tribes, most notably the Maasai. The Maasai are famous for their songs and flowery speech. The following common prayer, given at dawn by a Maasai *moran*, or warrior, is quoted by Saitoti and Beckwith:

> Greetings, heavenly dawn. You come to us in red and white. Our women compete to greet you. I come here, the blessed of the blessed, still resting beneath the tree. No wild animals, nor vultures' wings will tamper with us. No rhino horns or sharp spearpoints will separate us. I pray for prosperity. Let it come to us in slow, uphill motions. Let it come to stay. Wild beasts and vultures, hush—your expectations were not met, for we are all still alive. Bequeath us babies and cows, on the slopes and on the plains, when we search and when we don't. Give us prosperity by surprise. Keep us until the wrinkles of old age. Goodbye, heavenly dawn, until tomorrow when we shall meet again in peace and in the golden rays of prosperity.[28]

Spiritual Places

Places in nature have spiritual significance among many ethnic groups. The Maasai consider many trees, rocks, and mountains to be sacred. According to Tepilit Ole Saitoti in his book with Carol Beckwith titled *Maasai*,

"Whenever a Maasai passes by a holy tree, he or she will pluck green grass, put it on the tree, and then pray. Sometimes bead necklaces, bracelets, or anklets are placed on the tree as well."

The Kikuyu consider Mount Kenya to be sacred, the birthplace of the first Kikuyu, the first people, and the home of God, *Ngai*.

Prayers of thanksgiving (such as expressing gratitude for the daily sunrise) and requests for divine intervention (to solicit rain, for example) are common among many ethnic groups. Prayers are also offered during important cultural rituals and ceremonies. The Okiek, a Paranilotic ethnic group living in southwest Kenya, pray throughout ceremonies. Their prayers reveal the values most important to Okiek culture, most notably peace and harmony. As author Corinne Kratz reports in her book about the women of this tribe,

> Blessing punctuates Okiek ceremonies from beginning to end, and is an essential part of most ritual events. During initiation, elders pray for good fortune and peace all along the way, blessing people and the ceremony, ritual materials and the products made from them. . . .
>
> Each ceremonial blessing calls from peace, cooperation, and understanding, and reminds people to act accordingly, to create the harmonious

As done in many other religions, traditional East Africans offer prayers (pictured) for gratitude and to make requests.

atmosphere expected during ceremonies. What blessings pray for and how wishes are expressed in them reveal and create an image of ideal Okiek life. Peaceful, cooperative harmony is one of the four main themes and values expressed. . . . The others are fecundity [the bearing of children], prosperity and continuity.[29]

Cultures Mix in Religious Rituals

Though most ethnic groups, such as the Okiek, pray in their tribal languages for the preservation of their traditional values, others use different languages—and even different religions—to petition God for different aspects of their lives. This is common among tribes with traditional religious beliefs who have also been exposed to Islam or Christianity.

The Waso Boorana, part of the Oromo tribe of northern Kenya, are an example of an ethnic group that embraces the traditions of more than one religion. Outside their homes, the Waso Boorana are a devout Muslim people. They celebrate Muslim holidays and religious observances with the recitation of traditional Arabic prayers found in the Koran, the Muslim holy book, and visits to nearby mosques. Muslim prayers are also used in community-wide rituals to bless the entire Oromo tribe. At home, however, the Waso Boorana invoke traditional Oromo religious prayers and rituals. Their daily greeting, wishing others the "Peace of the Boorana," uses the Boorana language to express the highest ideal in Waso Boorana culture: peace.

The Boorana language is also used in the most important cultural ritual of the Waso Boorana, the *buna qalla*, or sacrifice of the coffee beans. The *buna qalla* ceremony, in which coffee beans are fried whole in butter and then eaten floating in milk or clarified butter, is, like the daily blessings in Boorana, "a part of a constant ritualized way of remembering Waso Boorana traditions and historical roots."[30] While the coffee beans are cooked, women bless their households and kitchens. While the coffee beans are eaten, elders invoke blessings and tell stories, anecdotes, and sayings to the children about the Waso Boorana's rituals and past. It is a way of connecting the Waso Boorana with God's blessings and their own history.

Despite their use of different languages and rituals to connect with different aspects of their community, the Waso

Waso Boorana Blessing

Some ethnic groups, like the Waso Boorana pastoralists in northern Kenya, have no form to their prayers. Elders improvise blessings depending on the needs of the situation. The following blessing was recorded by historian Mario I. Aguilar in his book *Being Oromo in Kenya*.

"ORIN NAGAA. WAT NAGAA, WATIYEN NAGAA, SAQ NAGAA, SAKUYE NAGAA, BOR NAGAA, BOORANI NAGAA, FULAN NU JIRCHUF NAGAA, BADA SADEN NAGAA, TULLA SALAN NAGAA, NAGENI HOLEGAL NAGAA BULEKAN, WAQ NAGAAN NUOLCH, WAQNUHORSIS.

LET IT BE PEACE!

let the animals live in peace [without being stolen, etc.]

let the Warta live in peace,

let the calves be in peace,

let Saq be in peace,

let the Sakuye have peace,

let Bor be in peace,

let the Boorana have peace,

let us be in peace wherever we stay,

let the *bada* [three sacred places of the Boorana in Ethiopia] be in peace,

let the nine spring wells be at peace,

let peace be with us throughout the day,

let peace be with us throughout the night,

may God grant us peace throughout the day,

may God make us rich [in terms of animals]."

Boorana believe in one God. According to anthropologist Mario Aguilar, "When asked about God, the Waso Boorana say that God is only one and that he speaks different languages. The God of the Boorana in Ethiopia, the God of Islam, and the God of Christianity are the same and the only one."[31]

Connecting with a divine force—by whatever name it is called—helps East Africans of all religions survive difficulties such as drought, disease, infertility, famine, or war. Many believe their difficulties are a result of some misdeed, sin, or broken taboo on the part of a tribe member.

Taboos and Curses

Taboos are strict cultural restrictions that, if broken, are believed to cause a variety of problems for individuals and groups. For example, among the Zaramo of coastal Tanzania, it is taboo for a father to see his daughter when she is under ritual seclusion during the weeks after she has first menstruated. It is believed that breaking this taboo can cause infertility in the girl when she marries and infertil-

A Maasai spiritual leader in Kenya. Spiritual beliefs give the Maasai and other cultures strength in difficult times.

ity in the people and fields of the community in general.

Broken taboos, however, are not believed to be the only cause of difficulties in traditional religions. Many East Africans believe that trouble can also be caused by curses from powerful people: witches, sorcerers, and those with the "evil eye" who curse because they mean harm to another.

Most East African ethnic groups, even those practicing Christianity or Islam, believe in the power of witches and those with the "evil eye." Witches and sorcerers, it is believed, cause others to fall ill by putting poison in their food or bewitching them. A Swahili villager who adheres to both Islam and traditional beliefs explained sorcery in this way to anthropologist Pat Caplan:

> Sorcerers kill by putting poison (*kibumbwi*) in food or drink—they keep it under their fingernails. They also know medicine to make themselves invisible, to fly and to become giants. And they can compel people to work for them at night, for example by sewing or driving a car. Sorcery is a matter of knowing the right medicine and one can go to a sorcerer and buy medicines. One can also be taught.

Another Swahili villager told Caplan about how a jealous person might cause harm to one who is more fortunate by using a curse called the evil eye:

> It is when someone says to him or herself, "Why does such a one do well and I don't get anything?" An evil eye means an evil heart. Someone has many children, or donkeys or cows. Someone sees that Mohammed has a fine child, Waziri, and he says, "Oh that God would take away this

Amulets

A shaman-diviner may offer protective words or even bless objects to be used as protection against evil. Such objects are called amulets. Often, parents will place amulets around the necks of their children to protect them from illness or other misfortune. German scientist Gunter Best describes the practice in his book *Marakwet and Turkana: New Perspectives on the Material Culture of East African Societies*.

"Youths and men also wear bead necklaces of one or more coils which often carry protective amulets . . . made of a branch and supposed to give protection from enemies and other persecutors. Women attach their amulets to their necklaces and men usually to a chain. . . . In dangerous situations a finger nail is [scratched] into the reverse side so that sawdust can be caught in the palm of the hand and blown by the person in danger in the direction of the pursuants. It is hoped that the pursuer can no longer see properly and gives up his pursuit."

Okiek Curse

The following curse, said during a boys' initiation ceremony in 1983, is intended in part to protect the boys going through circumcision. It was recorded by sociologist Corinne Kratz in her book about the Okiek titled *Affecting Performance*.

"Tell the spoiler

that bothers someone's child

Or sees that [a child] is bad

May the wind take you

May you go down with the moon

May you go down with the wind

If there is [someone] who knots [his] heart

If there is [someone] who knots [his] stomach

Oooooo

May the sun take you

Tell the kind people of cows, rise

To come and meet us when we come

To come and meet us when we come

Enemies

Tell them to keep away from us

Tell them may the wind take you

Tell them may the sticks on the path hurt you

May the light of dawn hate you."

child"—he has a bad heart, that is a thing of hatred (*chuki*).[32]

Elders and spiritual leaders may also curse someone to exact revenge for an evil deed. Curses of elders are very powerful, rarely used, and are one way in which an elder may administer justice with the help of a divine power. Anthropologist Paul Spencer questioned the Samburu on this matter:

> The Samburu say that the curse is like poison from a poison arrow tree: if it enters a place where the skin is cut (i.e. a wrong has been done) then it will inevitably kill, but if the skin is whole (i.e. no wrong has been done) then it will have no effect. The curse, then, unlike sorcery, cannot harm an entirely innocent person. If it is morally justified, however, it can lead to the most severe misfortune to the cursed man, his wives, his children or his cattle. God is the supreme arbiter who decides whether a curse is justified or not, and who brings misfortune to the wrong doer. It is he

> who confers a potent curse on certain persons. . . . And it is he who makes the curse of a mean man who is always resorting to it progressively less effective.[33]

Curses are only one way a tribe can handle the ill deeds of its members. Many ethnic groups believe in making up for mistakes such as broken taboos by performing certain rituals. For this process, called atonement, some tribes use animal sacrifices. One example comes from the practices of the Rendille, a Paranilotic pastoralist group in northern Kenya, "The Rendille have a blood sacrifice in their *soriu* ritual which is really an atonement for sin, for breaking taboos," reports Elliot Fratkin in an academic analysis of the region. "They mark their houses with the blood of the sacrificial lamb, just as the ancient Hebrews in Egypt did to protect against the angel of death."[34]

Many East African ethnic groups rely on spiritual leaders to lead them through the proper atonement rituals. Spiritual leaders play other roles, too, protecting individuals from evil and helping them find a cure for what ails them.

Some ethnic groups perform animal sacrifices to atone for mistakes. This Maasai warrior is eating an animal sacrificed from his herd.

Spiritual Leaders

In most ethnic groups, spiritual leaders are male elders. This is true in traditional East African religions, among Muslim clerics, and in many of the Christian churches of the region. Often these spiritual leaders also perform the role of judges in the culture, helping others resolve disputes. The *kabaka* of the Buganda people is an example of a spiritual leader who is also a judge and figurehead for the community, a symbol of the ideals of the Bugandan people.

But spiritual leaders are more than symbolic leaders. The people look to them for guidance to help overcome difficulties. According to Mario

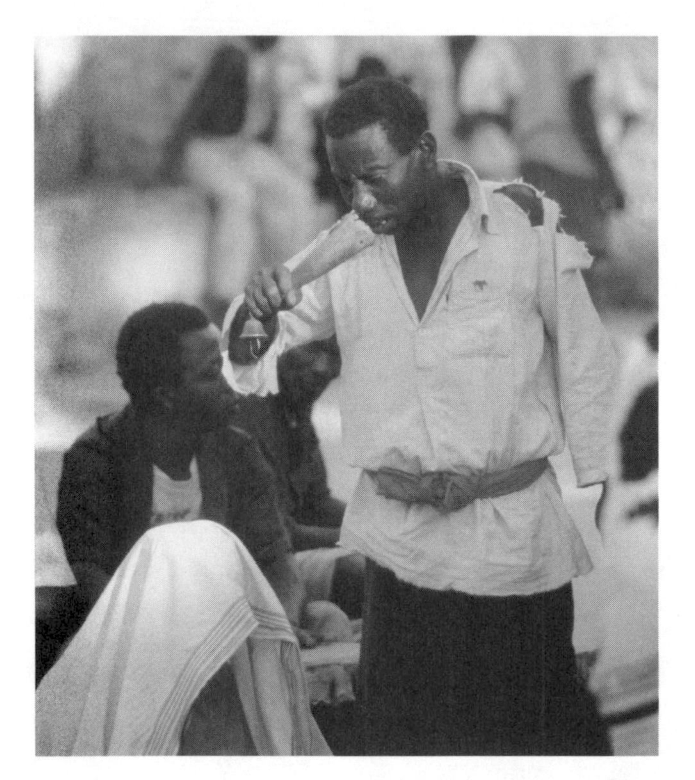

A Tanzanian spiritual leader performs an exorcism. Some religious leaders hold political as well as spiritual clout.

Aguilar, an expert on the Waso Boorana, "The ideal person to lead prayers or to bless or to settle disputes are those elders who know 'words', and can combine the instrumentality of those words with wisdom acquired through coping with the daily round."[35] The "words" these leaders know are often the words to traditional prayers. Some spiritual leaders know all the rituals associated with an ethnic group and preside over their clan's ceremonies throughout their lifetime. Among the Maasai and the Samburu, the most revered elder is the *laibon*, a man whose powers are at once spiritual and judicial, political and medical. One of his most important spiritual roles is that of diviner. Although among the Samburu and Maasai only a man can be a *laibon*, in some cultures women can be diviners as well.

Fulfilling the role of diviner—one believed to see the future and communicate with spirits—is an important part of a spiritual leader's work in many East African ethnic groups. The practice of divination also involves learning the causes of events in the present. Divination is practiced through the "reading" of stones, "reading" the entrails of sacrificed animals, and contacting helpful "spirits." The great *laibon* Simel of the Maasai allowed his experience of divination to be recorded in the book *Maasai*:

> Smooth river stones are placed in a calabash [a hollow gourd]; the diviner whispers and spits into the calabash, beseeching his great ancestors, Mbatian and Nelion, to aid him. He then pours out the stones, counting them out into small piles. Repeating the process a number of times, he will eventually make a prophecy or answer a question posed to him.[36]

Some spiritual leaders who practice divination call upon the spirit world to see into the future or answer questions. A Swahili villager who visited a traditional shaman-diviner, against the instructions of Islamic religious leaders, described his visit in this way:

You are sick, and someone calls Ali [the main shaman-diviner in the village]. He fumigates himself and tells

Women's Spirituality

Although men are most often the spiritual leaders in many ethnic groups, the women of the Maasai and some other tribes practice their own rituals and offer their own prayers when men are not present. Tepilit Ole Saitoti describes women's rituals in his book with Carol Beckwith titled *Maasai.*

"Whenever many Maasai women get together, they sing and dance among themselves. Beautifully adorned with multicolored bead necklaces and long, soft, lamb-hide dresses, they offer prayer songs, thanking God for His blessings or asking Him to bring them children, rain, grass, prosperity, and peace. Women in Maasailand pray to God more often than men do. As part of their prayer ritual, they sprinkle milk in three directions, the North, South, and East (the West, where the sun sets is used only for cursing). . . .

Women often gather in large prayer meetings, called *Alamal Loonkituak*, which are very emotional and exceedingly sentimental. All men, and particularly elders, are scared of the power of these meetings. Men may not do anything that might tamper with the meetings, but rather must supply all the things necessary for them, such as animals for sacrifices. Most of these large congregations gather to pray for children."

The following song is a song of blessing for a mother and her newborn child, reminding *Ngai* that prayer is constant.

*"Naomoni aaya*i.

The one who is prayed for and I also pray.

Naikurukur nesha,

God of the thunder and the rain,

Iye oshi ak-aaomon.

Thee I always pray.

Kileken oilepu,

Morning star which rises,

Iye oshi ak-aaomon.

Thee I always pray.

Paasai leleshway,

The Indescribable Color,

Iye oshi ak-aaomon.

Thee I always pray."

> his spirit that he is going to examine this person. When the patient arrives, he may do it again, calling on all the spirits by name. He is seated covered with cloths, and has his censer [incense burner] under the cloths with him. . . . Then he begins to sing *kitanga* songs, and the people who are there also sing and clap until he becomes possessed. Then he throws off the cloth and says, "What is going on here?" [The people there reply].[37]

The diviner may reveal, in the voice of the spirit that has "possessed" him, the reason for the illness or ill fortune, and may offer a solution or cure to the individual. The diviner's connection with the spirit world may be the most important part of his healing power.

The Spirit World

Many ethnic groups in East Africa, even those practicing Christianity or Islam, believe in the presence of spirits throughout the world. Often, these spirits are the spirits of ancestors who remain present to guide the tribe. It is said that the ancestor can both offer helpful advice and cause affliction. In *African Voices, African Lives*, author Pat Caplan describes these beliefs:

> Spirits can afflict humans for a variety of reasons, and in order for the affliction to cease, some kind of bargain will need to be made. Usually this consists of an offering of some kind, ranging from a few sweetmeats to the holding of a full-scale initiation ritual into a cult. Spirits which are known usually have shrines (*panga*) to which offerings can be taken, and if they are possessory they will have shamans (*wanganga*) who can control them, and mediums (*miti*) whom they possess at regular intervals. Some people are believed to have hereditary relationships (*asili*) with spirits, others are given to spirits as children to be "brought up" (*kulewa*) and thus protected. Sometimes spirits are thought to seek a relationship of friendship with a human, or to punish someone for annoying them.[38]

In some tribes, particularly Bantu-speaking tribes of Uganda, spirits are passed down from generation to generation.

Not all spirits are ancestors, however, or even human beings. Many ethnic groups believe that spirits inhabit animals, objects, or places in nature. The Waso Boorana culture is an example of this type of belief. Despite their adherence to Islam, the Waso Boorana have built a cult around the *riisi*, the spirit that resides in eagles and vultures. The Waso Boorana believe these birds to be messengers of God, Allah, or *Waqaa*, the Boorana word for God who resides in the sky. The Waso Boorana believe that when an eagle, or *riisi*, lands near a settlement, it is imperative for diviners to discover what it wants and what message it brings to the people of the settlement.

Spirits and spirit messengers are a common part of traditional African religions. They also provide a link between Islam

A Tanzanian Muslim man and boys pray. East African Islam has incorporated elements of traditional African religions.

and traditional African beliefs. Although Islam and Christianity generally discourage the belief in spirits as incompatible with monotheism, or the doctrine that there is only one God, Islam has been effective in drawing traditional beliefs into the Islamic worldview. Spirits are mentioned in the Koran as *jinn*, or genies, as they are known in the West. Many traditional African spirit cults have become associated with these *jinn*, thereby allowing Muslim East Africans to adhere to Islam as well as maintain many of their traditional beliefs.

Thus, despite generations of exposure to world religions such as Christianity and Islam, traditional African beliefs continue to affect the religious lives of many East Africans. Religious practices in the region have a decidedly African tone, with dancing, stories, drumming, and oratory. And religious beliefs continue to be affected by belief in the spirit world, a powerful connection with the forces of nature, and a reliance on a variety of religious leaders for guidance and intervention.

Chapter 6

East Africa Today: Problems and Possibilities

Present-day East Africa is a region of great contrasts. Tremendous poverty coexists with pockets of wealth. Modern hospitals exist in the same communities as traditional "witch" doctors. Rapid transit—trains, planes, and cars—intersects with people traveling long distances as they did in the Stone Age, walking barefoot. It is a place of crushing problems and fantastic hopes.

Some of the problems facing East Africans are indigenous—they come with the territory, so to speak. Health problems have always plagued the region. Though modern technology has made much of East Africa a healthier place to live, modern illnesses such as AIDS and the Ebola virus have taken a tremendous toll. Diseases such as malaria and cholera, and the environmental and economic nightmare of overpopulation, will exact much greater tolls before cures and solutions are found.

While health problems may come with the territory, not all of East Africa's problems are indigenous. Many of the region's most difficult problems were inherited from the colonial powers that granted them independence in the 1960s. Perhaps the most pressing problem, one that affects every aspect of life in East Africa, is economic underdevelopment, the result of the failure to use available resources to secure the potential benefits from the region's natural wealth. The economies of all three nations are unhealthily dependent on the developed nations of Europe, North America, and Asia, and they have not matured into self-reliance as a result of years of colonial economic exploitation.

Other unfortunate legacies of the years of European exploitation and colonization are political instability, violence, and repression. Tribalism—animosity and rivalry between tribes—encouraged in part by European colonial powers, continues to tear at the fragile fabric of nationhood and is the root of a great deal of the violence and instability that plagues East Africa. In Uganda especially, the struggle to create a sense of unity while maintaining the cultural integrity of more than a hundred different ethnic groups has torn the nation apart and caused violent political strife, which has often led to political repression. Moreover, political instability in countries outside East Africa creates pressures on the resources of the region as refugees flood in from the borders.

But East Africa is not a region without hope. The gifts of natural resources and a resourceful and increasingly well educated populace have worked wonders to increase the chances of a healthy and prosperous future for East Africa's residents.

Some Roots of Economic Underdevelopment

Perhaps the most pervasive problem that bars East Africa from achieving prosperity today is underdevelopment, the legacy of years of colonial rule. While Kenya, Uganda and Tanzania have taken different routes in their quest for prosperity, each nation continues to rely heavily on aid from developed nations for its most basic needs.

When they gained independence, Kenya, Tanzania, and Uganda inherited

Poverty is one of the problems faced by modern East Africa. Here, residents sit outside their ramshackle houses in a poverty-stricken town in Kenya.

economies that were oriented toward Europe and infrastructures—transportation, power, and water systems—that were in disrepair. In attempts to make the colonies pay for themselves, European nations had exploited their East African hosts and claimed the region's resources for Europe, as the editors of *Understanding Contemporary Africa* explain:

> Colonialism did not originate to assist African countries to develop economically. It originated to benefit European countries. That is not to say that African countries did not receive any benefits, but the growth or development that occurred in those countries was mainly peripheral to the growth and development of Europe. Only as it became clear that the colonies would soon seek independence did European countries begin to guide some of their colonies toward the goal of developing their own economies. . . .These efforts were minimal, however, and did little to make African economies self-sufficient. By the 1960's, the colonial administrations were being dismantled rapidly across the continent—but many economic ties to the former colonial powers remained.[39]

Today, Kenya, Uganda, and Tanzania are dependent not only on their former colonizers but on other wealthy industrialized nations such as the United States and Japan. In Kenya and Uganda, this dependency is in part the result of the emphasis Western nations place on industrial development over traditional agriculture. At the time of independence, the nations of East Africa primarily relied on farming to fuel their economies. At the encouragement of their colonial trading partners, Kenya and Uganda tried to turn away from their agricultural roots. National leaders believed that development could be best achieved by means of industrialization. Yet many problems resulted from this emphasis on industry over agriculture, and most remain to this day.

First, the focus on factories instead of farms forced both Kenya and Uganda into a debilitating state of debt. Politicians initially believed that building factories to produce goods once imported from Europe would save East Africa money on imports. This was not the case, however, because machines and materials to produce such goods had to be imported from developed nations at great expense. And because the focus on factories sometimes replaced the focus on farm exports, Kenya and Uganda lost the chance to earn the export income needed to pay for industrial machinery. With little export income, East African nations were forced to borrow money from developed countries in order to pay industrial development expenses. Such borrowing plunged East African nations into a state of debt with which they struggle today.

In addition, emphasizing industry took energy, investment dollars, and labor away from agriculture, the mainstay of most African economies. In Kenya and Uganda,

A textile factory in Dar es Salaam, Tanzania. An early emphasis on industry instead of agriculture after independence in Kenya and Uganda led to problems such as high unemployment and urban poverty.

people abandoned their farms and moved to the city to look for jobs in industry. Upon arriving, however, most were unable to find work because so few jobs were available. The migration from the farms led to problems such as high unemployment and urban poverty, low food production in the countryside—which made food imports necessary—and the collapse of agricultural exports. What agricultural exports remained tended to be of a single commodity, such as coffee in Uganda. And when the price of coffee dropped, income from exports, and what it could buy, plummeted. Uganda's president, Yoweri Museveni, illustrates Uganda's problems in his book *What Is Africa's Problem?*

> [Uganda's] economic misfortunes can be more graphically portrayed when you consider that in 1970, we needed 212 bags of coffee to buy a seven-ton Mercedes-Benz truck. In 1987 we required 420 bags and now we require more than 530 bags to secure the same vehicle.[40]

Ujamaa

While Kenya and Uganda were relying on industrialization and coffee plantation

Julius Nyerere's Freedom and Socialism: *Uhuru na Ujamaa*

Julius Nyerere, the first president of Tanzania, envisioned a society that differed greatly from that of most newly independent former colonies. Part of his social vision was the creation of Ujamaa villages, collective farming villages in which everyone worked and everyone benefited. He articulated his visionary plan for his nation in his 1968 book *Freedom and Socialism: Uhuru na Ujamaa.*

"UJAMAA AGRICULTURE

In a socialist Tanzania then, our agricultural organization would be predominantly that of co-operative living and working for the good of all. This means that most of our farming would be done by groups of people who live as a community and work as a community. They would live together in a village; they would farm together; market together; and undertake the provision of local services and small local requirements as a community. Their community would be the traditional family group, or any other group of people living according to ujamaa principles, large enough to take account of modern methods and the twentieth century needs of man. The land this community farmed would be called "our land" by all the members; the crops they produced on that land would be "our crops"; it would be "our shop" which provided individual members with the day-to-day necessities from outside; "our workshop" which made the bricks from which houses and other buildings were constructed, and so on. . . .

Such living and working . . . could transform our lives in Tanzania. We would not automatically become wealthy. . . . But most important of all, any increase in the amount of wealth we produce under this system would be "ours"; it would not belong to just one or two individuals, but to all those whose work had produced it. At the same time we should have strengthened our traditional equality and our traditional security. . . . For in each ujamaa village the man who is sick will be cared for; a man who is widowed will have no difficulty in getting his children looked after; the old, the unmarried, the orphans and other people in this kind of trouble will be looked after by the village as a whole, just as was done in traditional society."

In many ways, Nyerere's vision has come true. Most agriculturalists in Tanzania now live in Ujamaa villages within reach of school, health care, and government facilities. Collective farming has not been successful in providing Tanzania with food and export income, however. Private farms, privately owned and operated, now exist side by side with those owned by communities, profiting individuals as well as the nation as a whole.

agriculture for development, Tanzania was developing under an entirely different set of ideals. Julius Nyerere, the first president of Tanzania, believed that true independence could only be achieved if Tanzania learned to rely on itself. He called this model for independence *ujamaa*, a Kiswahili word that has been translated as "self-help," "mutual cooperation," and "familyhood." Nyerere felt that the cooperative nature of traditional East African agricultural villages could serve as a model for small-scale, self-reliant development and prosperity. Under "African socialism," as Nyerere's ideas have been labeled, Nyerere built new cooperative farming villages, called Ujamaa villages, with modern government services such as roads, schools, and health care clinics. Tanzanians of every ethnic stripe were invited—and sometimes forced—to move into these villages and farm surrounding lands together as a village collective. By 1974, most rural Tanzanians lived in Ujamaa villages.

Although it was hoped that Ujamaa villages would help Tanzanians become self-reliant, producing enough food to both feed their nation and provide export income, Tanzania has had as much trouble achieving economic independence as Kenya and Uganda have. Author Kevin Shillington outlines the situation:

Julius Nyerere, the first president of Tanzania and creator of the Ujamaa village system.

> On a national scale, Tanzania in the 1980s remained one of the poorest countries in Africa. It had huge foreign debts and was still dependent upon exporting agricultural raw materials—coffee, cotton, sisal—at prices controlled outside Africa, in exchange for increasingly expensive manufactured imports. But its production of food crops had not deteriorated, as in many other parts of Africa. It had avoided the massive accumulation of landless rural poverty which characterized its nationally more prosperous neighbour

An adult literacy class in an Ujamaa village in Tanzania in the 1970s. Education was part of Nyerere's vision for the farming collectives.

Kenya. And Tanzania had succeeded in providing the mass of rural people with vastly improved welfare services: clean water and free health and education facilities.[41]

Poverty and Hope

Tanzania, Kenya, and Uganda continue to be among the poorest nations in the world. And as the population of each nation has increased, its ability to feed its people and its per capita income, the annual income each citizen has to live on, have decreased. Hope abounds for East Africa's beleaguered economies, however, in a number of areas. As the twentieth century drew to a close, many developed nations promised aid in exchange for reforms in the political and economic arenas. In 1999, for example, international donors pledged nearly $1 billion toward Tanzanian development requirements in 2000. This aid was forthcoming in part because donors had noted encouraging signs that the government was stepping up its efforts to combat corruption.

Another area of hope is in the realm of development projects. Increasingly, creditors, those who lend money, are funding only those development projects that show local support, and grandiose plans are increasingly scaled back to encourage local management. The United Nations International Fund for Agricultural Development (IFAD), for example, lends money for infrastructural improvements such as water supplies, roads, and small-scale irrigation projects. IFAD loans also allow people too poor to own land to acquire some means of generating an income. Such projects emphasize the development of human resources as well as the conservation of natural resources.

Political Instability

A nation's ability to develop its human and natural resources often depends on its political openness and stability. All three East African nations have struggled to create politically stable, free, and fair societies in which human rights are respected.

Tanzania has been most successful in this arena. Despite occasional turbulence over Zanzibar's interest in independence, Tanzania has managed to hold a nation of diverse ethnicities together, ensuring free speech, fair elections, and smooth transitions of power. One reason for its success is a clear national vision and identity articulated by Tanzania's first president, Julius Nyerere, from the earliest days of independence. Through the national education system, each child is encouraged to see himself as Tanzanian rather than primarily as a member of an ethnic group. In addition, unlike many other African nations, Tanzania is increasingly ruled by laws and by well-balanced branches of government. This has been demonstrated by the smooth transitions of power through multiparty elections and by an independent judiciary that has successfully removed corrupt politicians from office.

Kenya and Uganda, however, have both struggled with political stability more than Tanzania has because of ethnic strife. In both nations, political parties tend to be divided along ethnic lines. Although multiple parties are now legal in Kenya, the voices of opposition groups, including politicians and the media, are routinely suppressed through intimidation campaigns supported by President Daniel arap Moi. According to a respected African survey,

> The Moi administration has consistently received international criticism of its record on human rights. . . . In July 1995, the United Kingdom withheld financial aid to Kenya, pending an improvement in the Moi administration's human rights policies and economic management.[42]

Surpassing Kenya in the violence of its political process is Uganda. Political instability caused by ethnic strife has frequently erupted into violence in the nation. In northern Uganda, a small-scale but continually violent armed struggle between a group of Acholi tribesmen called the Lord's Resistance Army and the Ugandan military has led to a life of terror for thousands of Ugandans since the late

The Lord's Resistance Army

Northern Uganda is the most violence-ridden area in East Africa. In the northern provinces of Gulu and Kitgum, a brutal rebel group, the Lord's Resistance Army, terrorizes the populace and battles with the Ugandan military. According to a 1997 report called *The Scars of Death*, human rights group Human Rights Watch reports,

"The rebel Lord's Resistance Army (LRA) is ostensibly dedicated to overthrowing the government of Uganda, but in practice the rebels appear to devote most of their time to attacks on the civilian population. . . .

When rebels move on, they leave behind the bodies of the dead. But after each raid, the rebels take away some of those who remain living. In particular, they take young children, often dragging them away from the dead bodies of their parents and siblings."

Boys and girls around fifteen or sixteen years old—and often as young as eight years old—are marched north to the rebel camps in the Sudan, where they are forced to fight as soldiers or labor as domestic slaves for the rebels. Those who try to escape are killed.

The Lord's Resistance Army began as an ethnic conflict between the mostly poor and uneducated Acholi people of northern Uganda, who would like to be in charge of the Ugandan government, and the more educated and wealthy Ugandans of the south who make up the majority of government officials. Starting in the late 1980s, rebel groups took on a religious tone and were led by Acholi prophetess Alice Lakwena and later her relative, Joseph Kony. One girl, Christine, who was abducted at age seventeen, recalls,

"The rebels call Joseph Kony their father, and say that the Holy Spirit speaks to him, and tells him what to do. . . . At times they pray like they're Christians, and at times like they're Muslims. . . . Their customs are strange. If they've just abducted you, they smear you with oil in the sign of the cross on your forehead and on your chest."

Despite the Ugandan military's efforts to eradicate the LRA, the rebels continue to plague northern Uganda. Although it is never strong enough to destablize the government, the Lord's Resistance Army causes great harm to the innocent people of the region—especially the children.

Lord's Resistance Army soldiers guard girls who were abducted from a school in northern Uganda in 1996.

Yoweri Museveni seized control of Uganda in 1986. Ten years later he became the country's first president.

1980s. But not all ethnic conflicts in Uganda have remained small in scale. Some have resulted in coups, which are sudden, forceful, and often violent transitions of power.

Coups and civil war from the 1970s to 1986 left the Ugandan people weary of conflict and hungering for fairness, peace, and prosperity. In 1986, Yoweri Museveni came to power, his National Resistance Army seizing the capital, Kampala. In 1996 he won Uganda's first presidential election in sixteen years. Museveni has struggled to bring political and economic stability to a region long torn by tribalism. He writes about the problems of tribalism in his book *What Is Africa's Problem?*

> In the National Resistance Movement [NRM] we ruthlessly oppose tribalism and the use of religion in politics. If you emphasize the interests of one tribe against those of other tribes, how can you build a nation? . . . How can you survive that way in the modern age?[43]

Despite his awareness of the problems of tribalism and the need to unify Uganda, Museveni has moved slowly in encouraging the shift to a more democratic and free society. As analyst Edward Bever sees it,

> Uganda under the NRM remains an undemocratic country in which the rights to choose the government, to be tried by an impartial court, to speak without fear of reprisal, to associate freely, and to organize independent labor unions have not yet been secured. It also remains one of the poorest countries in the world.[44]

Problems from Outside the Region

While most political crises are internal to the nations of East Africa and involve the struggle to create free and fair societies, some political problems come from outside the region. In fact, some political problems have flared up between the three nations. In the 1970s, for example, Uganda's dictatorial president Idi Amin put the country at war with both Kenya and Tanzania over borders and territory.

Other nations' problems have affected the region as well. Political and economic crises in Somalia, Ethiopia, Sudan, and Zaire have sent refugees fleeing into East Africa. In the 1990s, all three nations were impacted when ethnic strife between the Tutsi and Hutu tribes in Rwanda caused thousands of refugees to flood the borders of Kenya, Uganda, and Tanzania, putting a strain on overburdened resources.

Health Crises

The resources of East Africa, including human resources, have also been severely strained by the constant health crises that have beset the region. From relatively new problems such as AIDS, or acquired immune deficiency syndrome, and overpopulation to long-present diseases such as malaria and cholera, the ever-present challenge to East Africa lies in protecting people's health.

Education is one key to solving the region's number-one health and economic problem: overpopulation. At present rates, the population of East Africa will more than double in twenty years, outstripping the three national governments' abilities to feed and care for their people. But family-planning education shows promise in limiting population growth. In Kenya, education has helped lower the average birthrate from 7.8 children per mother in 1980 to 4.6 children per mother in 1995.

Education in the use of birth control devices such as condoms may also help prevent the spread of the newest of Africa's health problems, AIDS. In 1990, an estimated 10 million people worldwide carried HIV, the virus that causes AIDS. African nations, with 10 percent of the world's population, accounted for 25 to 50 percent of HIV infections. Today, the number of HIV carriers worldwide is upwards of 25 million people, over 75 percent of whom live in sub-Saharan Africa, Africa south of the Sahara Desert. East Africa is the heart of the so-called AIDS belt of Africa.

Women and children in East Africa have been particularly at risk of contracting the disease. Women are frequently infected by their husbands, who may have more than one wife or sexual partner. And children acquire HIV most often at birth, from infected mothers. Most tragically, because of the high incidence of AIDS in East Africa, the life expectancy, or average age to which people can expect to live, dropped to a low of forty-one years in Uganda in 1995.

AIDS has a particularly strong impact on the economy of East Africa as well, as explained in *Understanding Contemporary Africa*:

> The largest numbers of AIDS cases tend to be among men and women in

An AIDS education sign in Kampala, Uganda, one of the countries hit hardest by the AIDS epidemic.

> the most productive age groups. As the disease continues to spread, it could have a profound impact on the supply of labor, reducing the size and productivity of the labor force, including the highly trained sector. . . . In rural areas, there may be a reduction of the number of adults who can produce food.[45]

While AIDS and overpopulation provide a tragic backdrop to East Africa's health and economic problems, some hopeful rays of light are visible. Although the incidence of HIV infection continues to rise, education campaigns to teach East Africans family planning and safe sex practices may save millions of lives. In addition, countless babies are now being spared HIV infection because of new drugs that can be given to women during labor to prevent transmission of the virus. While neither of these factors will dramatically reduce the steep climb in AIDS cases in East Africa in the near future, even small efforts to save lives make an impact on the people of the region.

Children have also been positively impacted by another worldwide health campaign. Since the 1980s, immunization campaigns protecting babies against a multitude of diseases such as diphtheria, typhoid, cholera, whooping cough, and measles have been a priority for two agencies of the United Nations: the United Nations Children's Fund (UNICEF) and the World Health Organization (WHO). As a result, infant mortality rates have dropped in the region.

Still a Region of Contrasts

East Africa today continues to be a region of contrasts. For every good news report there seems to be a corresponding segment of bad news. And for every good report, there is a sobering incidence of difficulty. For example, while the infant death rate is falling, the adult death rate due to AIDS is climbing. And while the people celebrate political change and upcoming multiparty elections, govern-

Malaria

AIDS is not the only illness that threatens Africa's health and economic well-being. Each year, 300 million people worldwide, 90 percent of whom live in Africa, are sickened by malaria, a common, serious, and largely preventable tropical disease. One million people die annually, mainly young children living in remote regions of Africa without access to health services. As with other problems there, the poverty and lack of education of the people at risk pose serious obstacles to implementing solutions.

Malaria is transmitted when parasites are injected into the skin by mosquitoes, which bite between sunset and sunrise. Once in the human bloodstream, the parasites can cause damage to the liver and to other internal organs such as the brain by clumping together and blocking blood flow. According to the website of the World Health Organization, symptoms of malaria include the following:

"bouts of fever accompanied by other symptoms and alternating with periods of freedom from any feeling of illness. The intermittent type fever is usually absent at the beginning of the disease, when headache, malaise, fatigue, nausea, muscular pains, slight diarrhoea and slight increase of body temperature are the predominant and vague symptoms, often mistaken for influenza or gastro-intestinal infection. Most severe forms of the disease result in organ failure, delirium, impaired consciousness and generalized convulsions, followed by persistent coma and death."

In an effort to combat malaria throughout Africa, the United Nations designated April 25, 2001, as Africa Malaria Day, and began a campaign called Roll Back Malaria, which will feature concentrated efforts to increase the scope and effectiveness of treatment and prevention measures for Africans at risk. Priorities include a supply of affordable drugs, protective bed nets, and a focus on pregnant women and children.

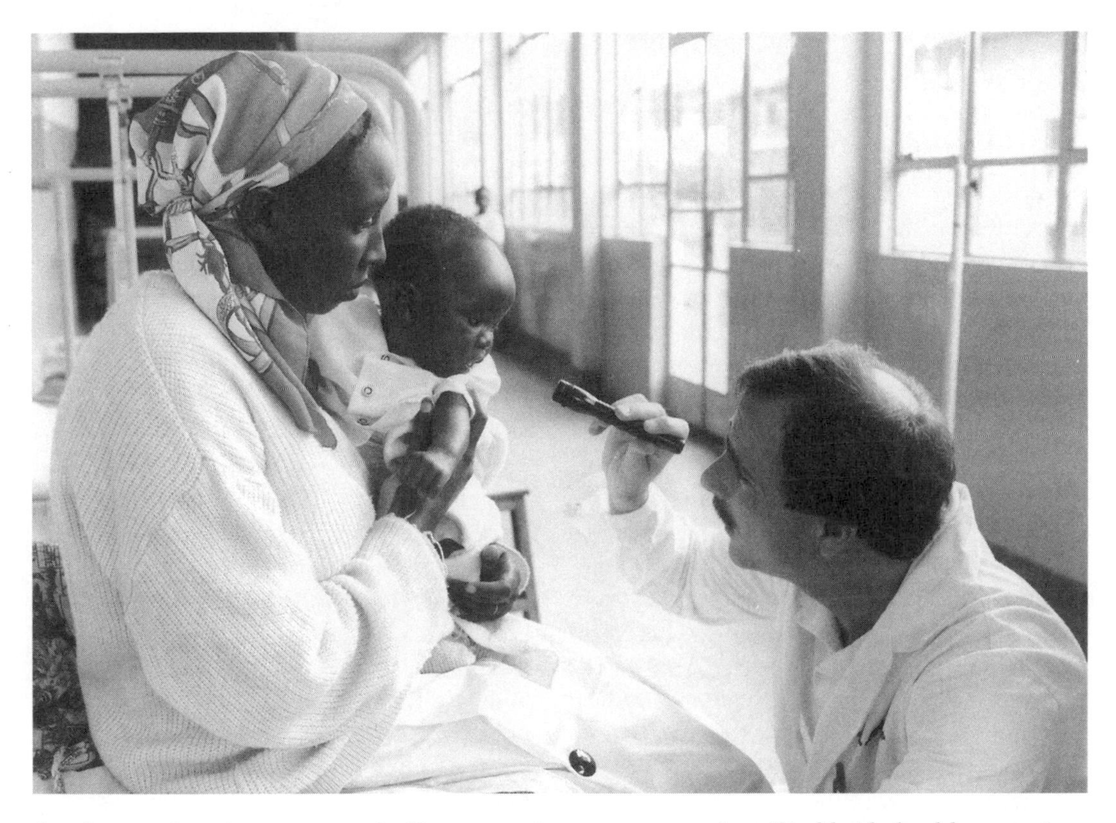

A volunteer American surgeon in Kenya examines a young patient. Worldwide health campaigns have greatly improved medical care for children in East Africa.

ments continue to suppress opposition parties.

But reason for hope abounds. United Nations officials are finally getting a handle on a wide range of childhood illnesses such as diarrhea, cholera, and diphtheria. Economic and political reforms have also been encouraged by another arm of the United Nations, the International Monetary Fund (IMF), which helps poor countries get their finances in order. In all three nations, the IMF and creditor nations have asked for reforms in exchange for funding. In Tanzania and Uganda, Presidents Benjamin Mkapa and Museveni have made efforts to reprivatize industries seized by the national governments in the 1970s and give their people a chance to thrive economically under their own efforts. And political anticorruption campaigns in all three countries have struggled to make inroads in creating governments that the people can trust.

East Africans of every ethnic group are working to make their governments just, their homes peaceful and prosperous, and their families healthy. In all that they do, the spirit of the people of

Corruption in East African Governments

Corruption has long been a problem among government officials and civil servants throughout Africa. Corruption has taken many forms. At high levels of government, officials may steal money from government coffers, often money that was loaned by international creditors to help in the nation's development. At lower levels of government, corruption often takes the form of bribery and graft, such as a policeman or other civil servant demanding money from citizens in return for favors. Throughout East Africa's postcolonial history, corruption often hindered the government's ability to serve their people. According to April and Donald Gordon, editors of *Understanding Contemporary Africa*,

"In many states, corrupt practices among strategically placed politicians and bureaucrats became so habitual as to be institutionalized. Under these circumstances, citizens expected to pay bribes; and they viewed politicians' raiding of government treasuries as simply 'the way things are done.'"

Today, anticorruption campaigns are having an impact throughout East Africa. Many creditor nations are refusing to lend money to corrupt governments, demanding reforms in exchange for loans. And many government leaders are responding to these demands. April and Donald Gordon document the progress of Africa's leaders:

"Unheard of a decade ago, Africa's rulers are acknowledging the urgent need for better leadership. . . . In May 1991, at the Kampala [Uganda] Forum on Security, Stability, Development, and Cooperation in Africa, delegates . . . agreed to the Kampala Document. . . . Its proposals aim to create a democratic and economically and politically integrated Africa with responsible and accountable leadership."

Although declarations such as the Kampala Document express African leaders' hopes to end official corruption, the East African people continue to struggle with corruption and the chaos it creates.

East Africa continues to rise to the challenges with which it is presented. Despite the difficulties in which they find themselves, East Africans always find reason for hope.

As Maasai women pray among themselves,

Enjoo iyook engeraa
Injoo iyook ingishu
Iye akekisilig

Give us children
Give us cattle
Our hope is with you always.[46]

Notes

Introduction: A Land of Contrasts

1. G. Mokhtar, ed., *General History of Africa. Vol. 2: Ancient Civilizations of Africa*. Paris: United Nations Educational, Scientific, and Cultural Organization, and Heinemann Educational Books, 1981, p. 583.

Chapter 1: Lifestyles of East African Ethnic Groups

2. Jomo Kenyatta, *Facing Mount Kenya*. London: Heinneman Educational Books, 1938, p. 3.
3. Tepilit Ole Saitoti and Carol Beckwith, *Maasai*. New York: Harry N. Abrams, 1980, p. 29.
4. Saitoti and Beckwith, *Maasai*, p. 204.
5. Marja-Liisa Swantz, *Ritual and Symbol in Transitional Zaramo Society*. New York: Africana Publishing, 1970, p. 88.
6. Mokhtar, *General History of Africa, vol. 2*, p. 571.

Chapter 2: The Arab Influence on East Africa

7. Quoted in Kevin Shillington, *History of Africa*. New York: St. Martin's Press, 1995, p. 124.
8. Quoted in Mokhtar, *General History of Africa, vol. 2*, p. 563.
9. Quoted in Mokhtar, *General History of Africa, vol. 2*, p. 561.
10. Shillington, *History of Africa*, p. 130.
11. Shillington, *History of Africa*, p. 211.
12. B. A. Ogot, ed., *General History of Africa. Vol. 5: Africa from the Sixteenth to Eighteenth Century*. Paris: United Nations Educational, Scientific, and Cultural Organization, and Heinemann Educational Books, 1989, p. 764.
13. Quoted in J. F. Ade Ajayi, ed., *General History of Africa. Vol. 6: Africa in the Nineteenth Century Until the 1880s*. Paris: United Nations Educational, Scientific, and Cultural Organization, and Heinemann Educational Books, 1989, p. 225.

Chapter 3: Europeans in East Africa

14. Shillington, *History of Africa*, p. 294.
15. Quoted in Shillington, *History of Africa*, p. 296.
16. Ludwig Von Hohnel, *Discovery of Lakes Rudolf and Stefanie: A Narrative of Count Samuel Teleki's Exploring and Hunting Expedition in Eastern Equatorial Africa in 1887 and 1888, vol. 1*. London: Frank Cass, 1968, pp. 1–2.
17. Shillington, *History of Africa*, pp. 238–39.

18. Shillington, *History of Africa*, p. 305.

19. Quoted in A. Adu Boahen, ed., *General History of Africa. Vol. 7: Africa Under Colonial Domination 1880–1935*. Paris: United Nations Educational, Scientific, and Cultural Organization, and Heinemann Educational Books, 1985, p. 397.

Chapter 4: Cultures of East African Ethnic Groups

20. Quoted in Ali A. Mazrui, ed., *General History of Africa. Vol. 8: Africa Since 1935*. Paris: United Nations Educational, Scientific, and Cultural Organization, and Heinemann Educational Books, 1993, p. 565.

21. Robert A. and Sarah LeVine et al., *Childcare and Culture: Lessons from Africa*. Cambridge, England: Cambridge University Press, 1994, pp. 152–53.

22. Saitoti and Beckwith, *Maasai*, p. 109.

23. Paul Spencer, *The Samburu*. Berkeley: University of California Press, 1965, p. 87.

24. Swantz, *Ritual and Symbol in Transitional Zaramo Society*, pp. 194–96.

25. LeVine et al., *Childcare and Culture*, p. 137.

26. Swantz, *Ritual and Symbol in Transitional Zaramo Society*, p. 199.

27. Pat Caplan, *African Voices, African Lives: Personal Narratives from a Swahili Village*. London: Routledge, 1997, pp. 91–92.

Chapter 5: Religion and Spirituality in East Africa

28. Quoted in Saitoti and Beckwith, *Maasai*, p. 121.

29. Corinne A. Kratz, *Affecting Performance: Meaning, Movement, and Experience in Okiek Women's Initiation*. Washington DC: Smithsonian Institution Press, 1994, pp. 165–67.

30. Mario I. Aguilar, *Being Oromo in Kenya*. Trenton, NJ: Africa World Press, 1998, p. 178.

31. Aguilar, *Being Oromo in Kenya*, p. 195.

32. Caplan, *African Voices, African Lives*, pp. 160–61.

33. Spencer, *The Samburu*, p. 186.

34. Elliot Fratkin, *Surviving Drought and Development: Ariaal Pastoralists of Northern Kenya*. Boulder, CO: Westview Press, 1991, p. 85.

35. Aguilar, *Being Oromo in Kenya*, p. 172.

36. Saitoti and Beckwith, Maasai, p. 261.

37. Quoted in Caplan, *African Voices, African Lives*, p. 158.

38. Caplan, *African Voices, African Lives*, p. 152.

Chapter 6: East Africa Today: Problems and Possibilities

39. April A. Gordon and Donald L. Gordon, eds., *Understanding Contemporary Africa*. Boulder, CO: Lynne Rienner Publishers, 1996, p. 95.

40. Yoweri K. Museveni, *What Is Africa's Problem?* Minneapolis: University of Minnesota Press, 2000, p. 211.

41. Shillington, *History of Africa*, p. 420.

42. J. A. Allen, et al., eds., *Africa South of the Sahara 2000*. London: Europa Publications, 2000, p. 585.
43. Museveni, *What Is Africa's Problem?* p. 13.
44. Edward Bever, *Africa: International Government and Politics Series*. Phoenix, AZ: Oryx Press, 1996, p. 155.
45. Gordon and Gordon, *Understanding Contemporary Africa*, p. 114.
46. Quoted in Saitoti and Beckwith, *Maasai*, p. 194.

For Further Reading

Books

Robert Barlas, *Cultures of the World: Uganda*. New York: Marshall Cavendish, 1993. Simple text and photos detail life in Uganda.

Laurel Corona, *Modern Nations of the World: Kenya*. San Diego, CA: Lucent Books, 2000. An intensive description and analysis of the modern nation of Kenya, with historical and cultural background.

Isak Dinesen, *Out of Africa*. New York: Modern Library, 1952. An account of a European woman's life in an East African plantation written under a pseudonym by Karen Von Blixen.

Amiram Gonen, ed., *Peoples of the World*. Danbury, CT: Grolier Educational, 1998. An encyclopedia of world peoples, with historical and current perspectives as well as anthropological and social data.

Robert Pateman, *Cultures of the World: Kenya*. New York: Marshall Cavendish, 1993. Simple text and bright photos detail life in a variety of cultures in Kenya.

Henry M. Stanley, *How I Found Livingstone: Travels, Adventures, and Discoveries in Central Africa*. London: Sampson Low, Marston, Searle, and Rivington, 1884. Famous explorer Stanley's illustrated, firsthand account of his adventures in search of Dr. Livingstone in the late 1800s.

Conrad R. Stein, *Enchantment of the World: Kenya*. Chicago: Children's Press, 1985. Comprehensive picture of the history, culture, politics, and economy of this East African nation.

Internet Sources

Children of East Africa (www.ccph.com/coea/). Developed by teachers, this "creative connections" site gives viewers a tour of an East African village led by two children, Chieng (Sunshine) and Atieno (Night).

Discover Kenya (www.afroam.org/children/discover/Kenya). This website offers comprehensive facts about the nation of Kenya in a youth-friendly format.

Integrated Regional Information Networks (www.reliefweb.int/irin/cea/ceafp). The United Nations Office for the Coordination of Humanitarian Affairs runs this website which offers up-to-date information on a variety of current issues in Kenya, Uganda, and Tanzania.

Kiswahili Home Page (http://conn.me.queensu.ca/kassim/documents/kiswa/swahil). Viewers can learn the history of the Kiswahili language and culture, also includes basic lessons in Kiswahili.

Moja (www.moja.com). Moja, meaning "one" in Kiswahili, is a website in English and Kiswahili that bills itself as "one newspaper and information source for East Africa—Kenya, Tanzania and Uganda."

Works Consulted

Books

Mario I. Aguilar, *Being Oromo in Kenya*. Trenton, NJ: Africa World Press, 1998. A description of the lives—particularly the spiritual lives—of the Waso Boorana people living in the arid northern reaches of Kenya.

J. A. Allen, et al., eds., *Africa South of the Sahara 2000*. London: Europa Publications, 2000. This nation-by-nation report on the politics, welfare, and economies of Africa is updated yearly.

Gunter Best, *Marakwet and Turkana: New Perspectives on the Material Culture of East African Societies*. Frankfurt, Germany: Museum fur Volkerkund, 1993. An academic analysis of the material culture—the things people use—of two pastoralist tribes in northern Kenya.

Edward Bever, *Africa: International Government and Politics Series*. Phoenix, AZ: Oryx Press, 1996. An account of the major problems facing Africa today.

Pat Caplan, *African Voices, African Lives: Personal Narratives from a Swahili Village*. London: Routledge, 1997. Anthropologist Pat Caplan lives in a Swahili village in Tanzania, recording and analyzing the views of local villagers on a variety of subjects.

Marshall S. Clough, *Mau Mau Memoirs: History, Memory and Politics*. Boulder, CO: L. Rienner, 1998. A historical analysis of the Mau Mau revolt in Kenya, including the firsthand accounts of survivors and participants.

Elliot Fratkin, *Surviving Drought and Development: Ariaal Pastoralists of Northern Kenya*. Boulder, CO: Westview Press, 1991. This academic study analyzes how pastoralists have survived political and ecological change in northern Kenya.

April A. Gordon and Donald L. Gordon, eds., *Understanding Con-*

temporary Africa. Boulder, CO: Lynne Rienner Publishers, 1996. Various writers offer a country-by-country report and analysis of contemporary Africa's politics, economies, and societies.

Human Rights Watch, *The Scars of Death: Children Abducted by the Lord's Resistance Army,* 1997.

Jomo Kenyatta, *Facing Mount Kenya*. London: Heinneman Educational Books, 1938. A history and cultural description of the Kikuyu people of Kenya written by the first president of that nation. Accessible to teenage readers.

Corinne A. Kratz, *Affecting Performance: Meaning, Movement, and Experience in Okiek Women's Initiation*. Washington, DC: Smithsonian Institution Press, 1994. An academic analysis of women's initiation practices among the Okiek people of East Africa.

Robert A. and Sarah LeVine et al., *Childcare and Culture: Lessons from Africa*. Cambridge, England: Cambridge University Press, 1994. This academic study offers an analysis of child-care practices among the Gusii, a Bantu-speaking ethnic group in Kenya, and compares Gusii child-care methods to American practices.

G. Mokhtar, et al., eds., *General History of Africa. Vols. 2–8*. Paris: United Nations Educational, Scientific, and Cultural Organization, and Heinemann Educational Books, 1981–1993. Produced through a massive effort by the United Nations, this multivolume series is seen by many as the most comprehensive work ever written on African history. Various authors report on and analyze the history and cultures of Africa using written historical records as well as archaeological, anthropological, political, economic, and linguistic data.

Yoweri K. Museveni, *What Is Africa's Problem?* Minneapolis: University of Minnesota Press, 2000. Uganda's leader since 1996 offers his views of the region's challenges and hopes in this collection of speeches.

Julius K. Nyerere, *Freedom and Socialism: Uhuru na Ujamaa*. London: Oxford University Press, 1968. A selection of political and philosophical writings and speeches from 1965 to 1967 by the first president of Tanzania.

John Roscoe, *The Baganda: An Account of Their Native Customs and Beliefs*. New York: Barnes and Noble, 1966. Roscoe's firsthand account of the Baganda (also called Buganda) around the turn of the twentieth century.

Tepilit Ole Saitoti, *The Worlds of a Maasai Warrior: An Autobiography*. New York: Random House, 1986. Saitoti's first-person account of his life in transition, from living as a traditional pastoralist to becoming a Harvard-educated leader. Accessible to teenage readers.

Tepilit Ole Saitoti and Carol Beckwith, *Maasai*. New York: Harry N. Abrams, 1980. This large-size book with vibrant glossy photographs and illustrations depicts all aspects of the lives of the Maasai people of Kenya and Tanzania from the perspective of a Maasai man. Accessible to teenage readers.

Kevin Shillington, *History of Africa*. New York: St. Martin's Press, 1995. A comprehensive history of Africa from prehistory to the present, in an entertaining and readable format.

Paul Spencer, *The Samburu*. Berkeley: University of California Press, 1965. This noted anthropologist and expert on pastoralist cultures offers a detailed description of the lives and culture of the Samburu Paranilotic pastoralists of Kenya.

Marja-Liisa Swantz, *Ritual and Symbol in Transitional Zaramo Society*. New York: Africana Publishing, 1970. An anthropological analysis of the spiritual lives of the Zaramo people, a Bantu-speaking fishing and farming culture in coastal Tanzania.

Ludwig Von Hohnel, *Discovery of Lakes Rudolf and Stefanie: A Narrative of Count Samuel Teleki's Exploring and Hunting Expedition in Eastern Equatorial Africa in 1887 and 1888*. Vol. 1. London: Frank Cass, 1968. This first-person, illustrated account of a famous expedition to East Africa offers descriptions of the region in the late 1800s. Accessible to teenage readers.

James Woodburn, *Hunters and Gatherers: The Material Culture of the Nomadic Hadza*. London: Trustees of the British Museum, 1970. Anthropologist James Woodburn offers this photo essay describing the artifacts of a disappearing culture of some of the last hunter-gatherers of East Africa.

Internet Sources

Amnesty International, "Female Genital Mutilation," 1997. www.amnesty.org.

World Health Organization, "Roll Back Malaria," 2001. www.who.int.

Index

Picture Credits

Cover photo: © CORBIS
© AFP/CORBIS, 91
© Paul Almasy/CORBIS, 38
© Adrian Arbib/CORBIS, 18
© Yann Arthus-Bertrand/CORBIS, 74, 77
© Bettmann/CORBIS, 43
© Bojan Brecelj/CORBIS, 41
© Jan Butchofsky-Houser/CORBIS, 25
© CORBIS, 9, 24, 57
© Eye Ubiquitous/CORBIS, 32
© Owen Franken/CORBIS, 33
© Dave G. Houser/CORBIS, 21, 30
Hulton/Archive by Getty Images, 36, 44, 45, 46, 52, 53, 87
© Hulton-Deutsch Collection/CORBIS, 49, 72
© Jason Lauré, 17, 59, 61, 95
© Jason Lauré 2000, 56
Lauré Communications, 85, 88
© Buddy Mays/CORBIS, 15, 65
© Joe McDonald/CORBIS, 20
© Jeffrey L. Rotman/CORBIS, 70
© Liba Taylor/CORBIS, 83, 90
© David & Peter Turnley/CORBIS, 66, 81
Werner Forman Archive/Art Resource, NY, 10
© K.M. Westermann/CORBIS, 23
© Nik Wheeler/CORBIS, 78
© Staffan Widstrand/CORBIS, 93

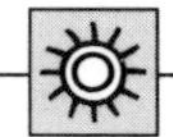

About the Author

Cynthia L. Jenson-Elliott first became interested in East Africa when studying African history as a college student at Bowdoin College in Maine. In 1984, she spent five months in Kenya and Tanzania through the St. Lawrence University Semester in Kenya. During that time she lived with a variety of ethnic groups in Nairobi and rural areas throughout Kenya, including the Luhya, Samburu, and Gabbra people. Ms. Jenson-Elliott holds a master's degree in education and has worked as a teacher, environmental and museum educator, and educational writer. She is currently a stay-at-home mom. She is also the coauthor of a children's book on cheetahs for the Zoobooks series. This is her first book for middle school students.